THE COMPLETE GUIDE TO MIDDLE SCHOOL MATH

GRADES 6-8

AMERICAN MATH ACADEMY

By H. TONG, M.Ed.

Math Instructor & Olympiad Coach

www.americanmathacademy.com

THE COMPLETE GUIDE TO MIDDLE SCHOOL MATH
GRADES 6-8

Writer: H.Tong
Copyright © 2020 The American Math Academy LLC.

All rights reserved. No part of this publication may be reproduced in whole or in part, stored in a retrieval system, or transmitted in any form or by any means electronic, mechanical, photocopying, recording or otherwise, without written permission of the copyright owner.

Printed in United States of America.

ISBN: 9798870973944

Although the writer has made every effort to ensure the accuracy and completeness of information contained this book, the writer assumes no responsibility for errors, inaccuracies, omissions or any inconsistency herein. Any slighting of people, places, or organizations is unintentional.

Questions, suggestions, or comments, please email: americanmathacademy@gmail.com

TABLE OF CONTENTS

CHAPTER I ARITHMETIC

CHAPTER II ALGEBRA

CHAPTER III GEOMETRY

CHAPTER IV PROBABILITY & STATISTICS

About the Author

Mr. Tong teaches at various private and public schools in both New York and New Jersey. In conjunction with his teaching, Mr. Tong developed his own private tutoring company. His company developed a unique way of ensuring his students' success on the math section of the ACT/SAT. His students, over the years, have been able to apply the knowledge and skills they learned during their tutoring sessions in college and beyond. Mr. Tong's academic accolades make him the best candidate to teach ACT/SAT Math. He received his master's degree in Math Education. He has won several national and state championships in various math competitions and has taken his team to victory in the Olympiads. He has trained students for Math Counts, American Math Competition (AMC), Harvard MIT Math Tournament, Princeton Math Contest, National Math League, and many other events. His teaching style ensures his students' success. He personally invests energy and time into his students and sees what they're struggling with. His dedication to his students is evident through his students' achievements.

Acknowledgements

I would like to take the time to acknowledge the help and support of my beloved wife, my colleagues, and my students, their feedback on this book was invaluable. Without their help, this book would not be the same. I dedicate this book to my precious daughters Vera, and Nora who were my inspiration to take on this project.

THE NUMBER SYSTEM

Whole Numbers: A number without a decimal or fraction

Examples: 0, 1, 2, 3, 4, 5, 6, 7, 8, 9…

Study Tips:

- ✓ Whole numbers are the numbers 0, 1, 2, 3, 4, 5, 6, 7 and so on
- ✓ 0 is the smallest whole number
- ✓ Negative numbers are not considered whole numbers.

Natural Numbers: The set of counting numbers, starting at 1 and going up.

Examples: 1, 2, 3, 4, 5, 6, 7, 8, 9…

Study Tips:

- ✓ Natural numbers are the numbers 1, 2, 3, 4, 5, 6, 7 and so on.
- ✓ 1 is the smallest natural number.
- ✓ Negative numbers are not considered natural numbers.

Integers: The set of all whole numbers from positive to negative infinity.

Examples: ...−6, −5, −4, −3, −2, −1, 0, 1, 2, 3, 4…

Positive Integers: Whole numbers greater than zero.

Examples: 1, 2, 3, 4, 5, 6, 7…

Negative Integers: Whole numbers less than zero.

Examples: −1, −2, −3, −4, −5, −6, −7…

Opposites: Numbers that are the same distance from zero but on different sides of zero.

Example: The opposite of 4 is −4

Example: The opposite of −5 is 5

Example: The opposite of 0.6 is −0.6

Example: The opposite of $\frac{1}{2}$ is $-\frac{1}{2}$

Example: The opposite of x is −x

Reciprocal: The reciprocal is the multiplicative inverse of a number.

Example: The reciprocal of $\frac{x}{y}$ is $\frac{y}{x}$

Example: The reciprocal of $\frac{1}{2}$ is 2

Example: The reciprocal of $\frac{3}{4}$ is $\frac{4}{3}$

Example: The reciprocal of 7 is $\frac{1}{7}$

THE NUMBER SYSTEM

Rational numbers: A rational number is a number that can be in the form of $\frac{A}{B}$ where A is not equal to zero. ($B \neq 0$)

Example: $\frac{1}{2}$

Example: $2 = \frac{2}{1}$

Example: $0.5 = \frac{5}{10} = \frac{1}{2}$

Example: $-10 = \frac{-10}{1}$

Example: $0.777777... = 0.\overline{7}$

Study Tips:

- ✓ Every whole number is a rational number
- ✓ Every natural number is a rational number
- ✓ Every integer is a rational number
- ✓ Every repeating decimal is a rational number

Irrational Numbers: Any real number that cannot be written in fraction form or has non–repeating decimals.

Example: $\pi = 3.1415...$

Example: $1.2345...$

Example: $\sqrt{45} = 6.70820...$

THE NUMBER SYSTEM

Study Tips:

✓ If a square root is not a perfect square, then it is considered an irrational number

✓ Irrational numbers have no exact decimal equivalents

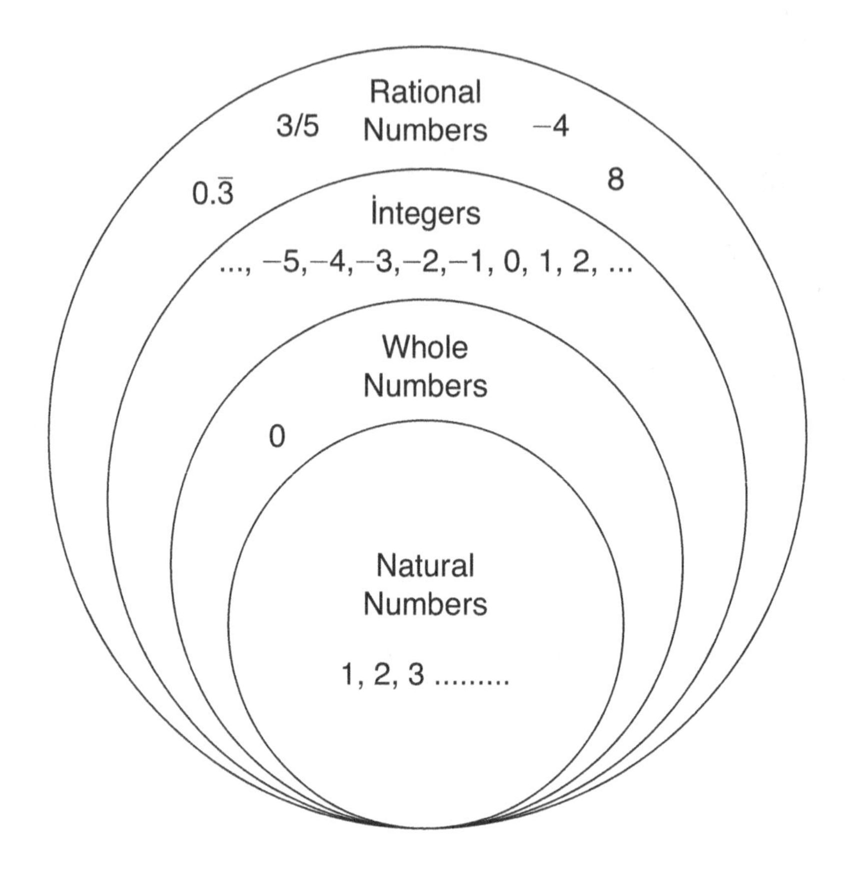

Number System Worksheet

1. **Directions:** Write each number of the following set in the correct location on the Venn diagram of the real numbers. Write each number only once.

$(-12, 3.75, 5, \sqrt{10}, \frac{3}{5}, 0, \sqrt{25}, \pi, \overline{3})$

Real Numbers	
Rational Numbers	Irrational Numbers
Integers Whole Numbers	

2. **True or false?** If false, explain why.

A) All whole numbers are rational numbers

B) All integers are rational numbers.

C) Some integer are irrational numbers.

Number System Worksheet

3. For each number shown, classify it as either rational or irrational.

	Rational or Irrational?
0	
$-\frac{1}{2}$	
$\sqrt{20}$	
$\sqrt{144}$	
$4\frac{1}{3}$	
0.35	
$\overline{9}$	
–9	
4.5	
$\sqrt{7}$	

4. Find the reciprocal for each of the following

A) $\frac{3}{5}$ B) 10 C) -5 D) 1

Number System Challenge

1. Which of the following numbers is irrational?

A) 3 B) -5 C) $\sqrt{7}$ D) $\sqrt{25}$

2. Which of the following numbers is rational?

A) $\sqrt{15}$ B) π C) 1.2345... D) $\frac{3}{5}$

3. Which of the following numbers is the opposite number of $-\frac{2}{3}$?

A) -2 B) -3 C) $-\frac{2}{3}$ D) $\frac{2}{3}$

4. Which of the following numbers is the opposite number of $-(-4)$?

A) 4 B) -4 C) $-\frac{1}{4}$ D) $\frac{1}{4}$

5. Which of the following numbers is a whole number but not a natural number?

A) -1 B) 0 C) 1 D) 2

6. Which of the following numbers is the reciprocal of $\frac{8}{11}$?

A) 8 B) 11 C) $\frac{11}{8}$ D) $-\frac{11}{8}$

7. Which of the following numbers is the reciprocal of 23?

A) 23 B) $\frac{1}{23}$ C) -23 D) 0

8. Which of the following is the reciprocal of $\frac{a}{b}$?

A) $\frac{a}{b}$ B) $\frac{b}{a}$ C) $-\frac{a}{b}$ D) $-\frac{b}{a}$

9. Which of the following number is the reciprocal of $\frac{1}{3} \cdot \frac{3}{6}$?

A) $\frac{1}{18}$ B) $\frac{1}{6}$ C) 6 D) -6

ORDER OF OPERATIONS

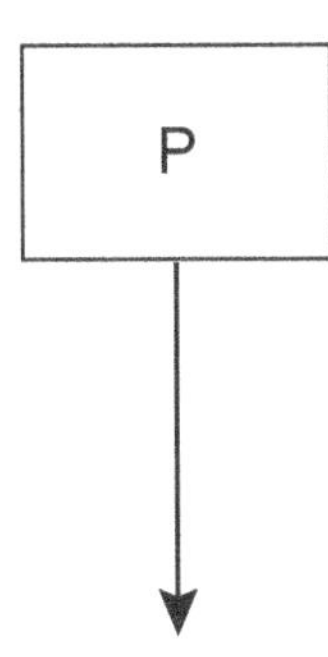

Parentheses

Do any operations in parentheses

$36 + 4(8 - 3) - 2^3 \div 4$

$36 + 4 \times 5 - 2^3 \div 4$

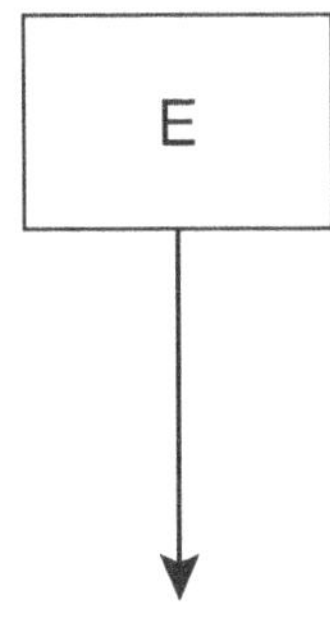

Exponents

Do any Exponents

$36 + 4 \times 5 - 2^3 \div 4$

$36 + 4 \times 5 - 8 \div 4$

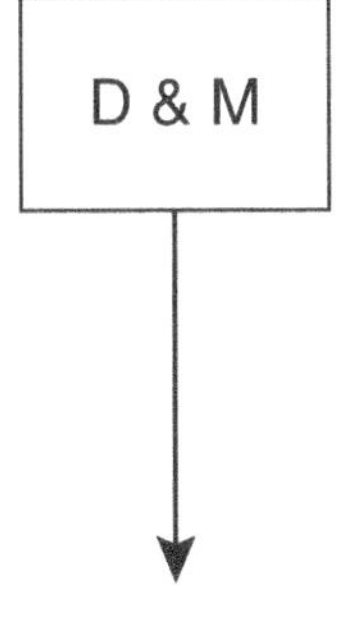

Divide and multiply rank equally (Always work, salve, or move, etc left to right)

$36 + 4 \times 5 - 8 \div 4$

$36 + 20 - 2$

A & S

Add and subtract rank equally (Always work, salve, or move, etc left to right)

$36 + 20 - 2$

$56 - 2$

54

Order of Operations Worksheet

Direction: Solve the following problems with **PEMDAS**

1. $3 + 2^2 \times 6 - 5$

2. $4^2 \times 6(7 - 3) - 5$

3. $3 + 2^0 \times 6 - 5$

4. $1 - 15 \times 4 + 16 - 6$

5. $(8 + 4) \times 5 \div 15 - (9 - 5)$

6. $3 \times (4 \times 3 + 15) - 8$

7. $(2 + 5)^2 - (13 + 9 \div 3)$

8. $3(5 \times 6 - 4) + 3^3$

9. $8 \times 9 + 2(15 + 4) - 42 \div 6$

10. $\frac{9}{4-1}(5 + 3) - 8$

Order of Operations Challenge

1. Which of the following is equal to $4(3 - 3) + 5^2$?

A) 0 B) 15 C) 20 D) 25

2. Which of the following is equal to $7(2 \times 3 - 4) + 4^3$?

A) 48 B) 54 C) 78 D) 108

3. Which of the following is equal to $\frac{14}{3-1}(7 - 3) - 12$?

A) 16 B) 20 C) 24 D) 36

4. Which of the following is equal to $4(2x + 5)$?

A) $2x + 5$ B) $4x + 20$ C) $8x + 5$ D) $8x + 20$

5. Which of following algebraic equations correctly represents this sentence:
Thirty–four is four times a number, increased by nine.

A) $34 = 4x - 9$ B) $34 = 4x + 9$ C) $4 = 9x + 34$ D) $9 = 34x + 4$

Order of Operations Challenge

6. Which of the following is equal to $19 - 5 + 3^3$?

A) 14 B) 27 C) 41 D) 51

7. Which of the following is equal to $54 - 4 \times 12 + 8$?

A) −12 B) −14 C) 12 D) 14

8. Which of the following is equal to $4 \times 5 - (6+3) - 63 \div 7$?

A) 0 B) 2 C) −4 D) −6

9. Which of the following number sentences is a correct match with the following sentence: 15 greater than the product of 9 and 11.

A) $9 \times 11 - 15$ B) $9 \times 11 + 15$ C) $15 \times 11 - 9$ D) $15 \times 11 + 9$

10. Which of the following equations matchs the following sentence:
Subtract 36 from 48 and then divide by 3

A) $(48 - 36) - 3$ B) $(48 - 36) + 3$ C) $(48 - 36) \div 3$ D) $(36 - 48) \div 3$

PRIME & COMPOSITE NUMBERS

Prime Numbers	Composite Numbers
A number that has only two factors, 1 and itself. 2, 3, 5, 7, 11, 13, 17, ...	A number that has more than two factors 4, 6, 8, 10, 12, 14, ...

0 and 1 are neither

Prime & Composite Numbers Worksheet

1. Circle the numbers that are prime numbers:

1	2	3	4	5
6	7	8	9	10
11	12	13	14	15
16	17	18	19	20
21	22	23	24	25

2. Circle the numbers that are composite numbers:

1	2	3	4	5
6	7	8	9	10
11	12	13	14	15
16	17	18	19	20
21	22	23	24	25

Prime & Composite Numbers Challenge

1. Which of the following is a prime number?

A) 22 B) 28 C) 32 D) 37

2. Which of the following is a composite number?

A) 13 B) 47 C) 53 D) 65

3. Mr. Johnson gave the following list of numbers to his class. He asked the class to find all of the composite numbers in the list. 3, 4, 7, 11, 14, 19, 21, 33 Which of these shows all of the composite numbers in the list?

A) 3, 5, 11, 19 B) 4, 14, 21, 33 C) 4, 19, 21, 33 D) 4, 11, 21, 33

4. If x is a prime number, which of the following could also be a prime number?

A) $\frac{x}{2}$ B) $x - 2$ C) $2x$ D) x^2

5. Which of the following numbers has more than two factors?

A) 11 B) 13 C) 15 D) 17

6. Which of the following is not a composite number?

A) 12 B) 13 C) 14 D) 15

7. Which of the following is an even prime number?

A) 2 B) 5 C) 7 D) 16

8. Which of the following is an odd composite number?

A) 1 B) 2 C) 6 D) 9

9. Which of the following is not a composite number?

A) 2 B) 4 C) 6 D) 8

10. Which of the following is the smallest prime number?

A) 1 B) 2 C) 3 D) 5

DIVISIBILITY RULES

Divisibility Rules		
Number	**Rules**	**Example**
2	A number is divisible by 2 when the last digit is even.	324 is divisible by 2 because the last digit is even.
3	A number is divisible by 3 when the sum of the digits is divisible by 3.	348 is divisible by 3 because 3+4+8=15 which is divisible by 3.
4	A number is divisible by 4 when the last two digits of the number are divisible by 4.	124 is divisible by 4 because the last two digits of the number are divisible by 4.
5	A number is divisible by 5 when the last digit of a number is 0 or 5.	675 is divisible by 5 because the last digit of the number is 5.
6	A number is divisible by 6 when the number is divisible by both 2 and 3.	336 is divisible by 6 because the number is divisible by both 2 and 3.
8	A number is divisible by 8 when the last 3 digits form a number that is divisible by 8.	888 is divisible by 8 because last 3 digits form a number that is divisible by 8.
9	A number is divisible by 9 when the sum of the digits of the number is divisible by 9.	981 is divisible by 9 because the sum of the digits of the number is divisible by 9
10	A number is divisible by 10 when the number ends in 0.	1200 is divisible by 10 because the number ends in 0.

Divisibility Rules Worksheet

Evaluate if the number divisible by each number. (Yes or No)		
1. $13 \div 2 =$	$108 \div 8 =$	$1000 \div 5 =$
2. $123 \div 3 =$	$189 \div 9 =$	$346 \div 3 =$
3. $124 \div 4 =$	$112444 \div 3 =$	$12348 \div 3 =$
4. $125 \div 5 =$	$1200 \div 10 =$	$1289 \div 9 =$
5. $72 \div 6 =$	$12 \div 8 =$	$866 \div 6 =$
6. $777 \div 7 =$	$349 \div 6 =$	$168 \div 7 =$

Divisibility Rules Challenge

1. Which of the following numbers is divisible by 1, 3, and 9?

A) 103 B) 108 C) 123 D) 148

2. Which of the following can be the missing digit of the following number that is divisible by 6.

848?96

A) 2 B) 4 C) 8 D) 9

3. Which of the following numbers is the smallest positive integer which when divided by 3 and 9 leave a remainder of 1?

A) 103 B) 105 C) 107 D) 109

4. Which of the following numbers is divisible by 1, 2, 5, and 10?

A) 18 B) 32 C) 45 D) 50

5. Which of the following numbers is divisible by 5?

A) 1 B) 15 C) 17 D) 24

Divisibility Rules Challenge

6. Which of the following numbers is divisible by 6?

A) 14 B) 16 C) 18 D) 20

7. Which of the following numbers is divisible by 10?

A) 9 B) 22 C) 33 D) 60

8. 102 is divisible by which following number?

A) 3 B) 5 C) 7 D) 9

9. There are 150 students in Science Olympiad competition. Each team is to have the same number of students. Can each team have 2, 3, 5 or 6 students?

A) Yes B) No

10. Which one of the following numbers is divisible by both 2 and 5?

A) 24 B) 36 C) 58 D) 60

LEAST COMMON MULTIPLE & GREATEST COMMON FACTOR

Factors: A fractor is a number that will divide into another number without a remainder.

Example: $42 \div 6 = 7$

Prime Factorization: The form of a number written as the product of its prime factors.

Example: The prime factorization of 96:

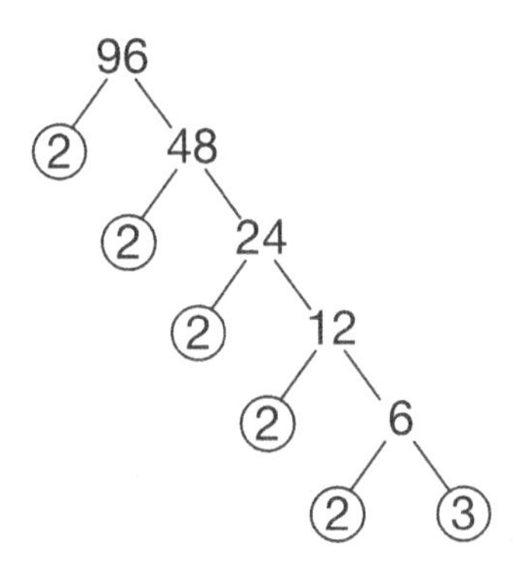

The prime factorization of 96 is $2 \cdot 2 \cdot 2 \cdot 2 \cdot 2 \cdot 3 = 2^5 \cdot 3$

Least Common Multiple (LCM): The least common multiple of two numbers is the smallest integer that is a multiple of both numbers.

Example: Find the LCM of 15 and 20.

Solution:

Multiples of 15: 0, 15, 30, 45, **60**, ...

Multiples of 20: 0, 20, 40, **60**, ...

60 is the least common multiple of 15 and 20.

Greatest Common Factor (GCF): The largest number that is a factor of two or more numbers.

Example: Find the GCF of 10 and 15.

Solution:

Factors of 10: 1, 2, **5**, 10.

Factors of 15: 1, 3, **5**, 15

5 is the greatest common factor of 10 and 15.

LCM & GCF Worksheet

Find the greatest common factor (GCF) and least common multiple (LCM) of each pair of numbers

	GCF	LCM
15, 45		
16, 24		
18, 48		
5, 45		
11, 88		
5, 10, 15		
16, 24, 36		

LCM & GCF Challenge

1. What is the least common multiple of 4 and 9?

A) 4 B) 9 C) 36 D) 72

2. What is the least common multiple of 7 and 8?

A) 14 B) 21 C) 28 D) 56

3. What is the Greatest Common Factor of 7 and 21?

A) 1 B) 7 C) 14 D) 21

4. What is the greatest common factor of 16 and 28?

A) 4 B) 14 C) 16 D) 56

5. What is the least common multiple of 3, 6 and 9?

A) 6 B) 9 C) 12 D) 18

6. What is the greatest common factor of 16, 18 and 36?

A) 2 B) 4 C) 6 D) 8

7. If x is the greatest prime factor of 21 and y is the greatest prime factor of 57, what is the value of $x + y$?

A) 12 B) 15 C) 26 D) 30

8. $A = 3x + 1 = 4y + 2 = 5z + 3$

From the above equations x, y, and z are positive integers and A is a two digit number. What is the smallest value of A?

A) 58 B) 68 C) 78 D) 88

ABSOLUTE VALUE

Absolute Value: The distance of integers from zero on the number line.

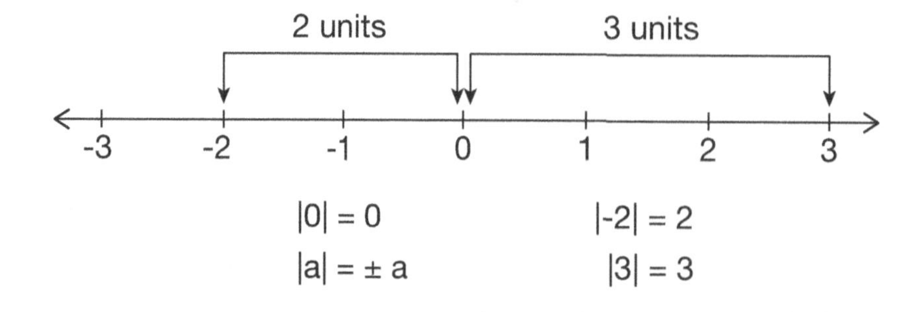

① if $a > 0$ then $|a| = a$

② if $a < 0$ then $|a| = -a$

Inequality Symbols	Number Line Symbols
$\leq$	←——●
$\geq$	●——→
$<$	←——○
$>$	○——→

Absolute Value Worksheet

Find the absolute values.

1. $|-5|$ ______________

2. $|-12|$ ______________

3. $|-24|$ ______________

4. $|97|$ ______________

5. $|106|$ ______________

6. $|17|$ ______________

7. $|-2|$ ______________

8. $|77|$ ______________

9. $|-8|$ ______________

10. $|25|$ ______________

Compare. Use <, >, or = .

11. $|-7|$ ______________ $|-5|$

12. $|39|$ ______________ $|-28|$

13. $|40|$ ______________ $|-28|$

14. $|82|$ ______________ $|75|$

15. $|-128|$ ______________ $|128|$

16. $|15|$ ______________ $|-12|$

17. 69 ______________ $|-69|$

18. $|-8|$ ______________ 0

19. $|14|$ ______________ 19

20. $|3|$ ______________ $|-3|$

Absolute Value Challenge

1. Find the value of the following expression.

$-5\,|4^2 - 8| - 12 = ?$

A) -12 B) -36 C) -42 D) -52

2. Evaluate the following using $x = 3$ and $y = 4$:

$|2x^2 - 5y| = ?$

A) -2 B) 2 C) 4 D) 6

3. Evaluate the following using $x = -12$ and $y = 6$:

$\left|\frac{2x}{y}\right| = ?$

A) -2 B) -4 C) 2 D) 4

4. Evaluate $|2^2 + (-3)^2|$.

A) -5 B) -1 C) 1 D) 13

Absolute Value Challenge

5. $\frac{|-9|+|5|}{|-7|}$ is equal to which of the following?

A) 1 B) 2 C) 5 D) 9

6. If $x < 0$ then $\frac{|x|+|-2x|}{|-3x|}$ is equal to which of the following?

A) -1 B) 0 C) 1 D) 2

7. Which statement is modeled by $3x + 7 > 10$?

A) The sum of 7 and 3 times x is at most 10.

B) Seven added to the product of 3 and x is greater than 10.

C) Three times x plus 7 is at least 10.

D) The product of 3 and x added to 7 is 10.

8. For the following inequality, which of the following can be x?

$|3x - 5| = 2x$

A) -5 B) -3 C) -1 D) 1

FRACTIONS & TYPE OF FRACTIONS

Fractions: Fractions are numbers that can be in the form A/B where B is not equal to zero.

Examples: $\frac{1}{5}, \frac{1}{7}, \frac{1}{8}, \ldots$

Types of Fractions

Proper fractions: A fraction where the numerator is less than the denominator.

$$\left.\begin{array}{l} A \longrightarrow \text{numerator} \\ B \longrightarrow \text{denominator} \end{array}\right\} A<B$$

Example: $\frac{1}{3}$

Improper fractions: A fraction where the denominator is less than the numerator.

$$\left.\begin{array}{l} A \longrightarrow \text{numerator} \\ B \longrightarrow \text{denominator} \end{array}\right\} A>B$$

Example: $\frac{8}{4}$

Mixed Fractions: When a fraction is written in the form $A\frac{B}{C}$

Example: $1\frac{1}{4}$

Operations with Fractions

Adding Fractions: When you add fractions, if they have same denominator, you add the numerators while keeping the denominator the same.

If the fractions have different denominators:

✓ Find the smallest multiple (LCD) of both numbers.

✓ Rewrite the fractions as equivalent fractions with the LCD as the denominator.

Key: $\frac{a}{b}+\frac{c}{b}=\frac{a+c}{b}$

$$\frac{a}{b}+\frac{c}{d}=\frac{a\cdot(d)}{b\cdot(d)}+\frac{c\cdot(b)}{d\cdot(b)}=\frac{ad+cb}{bd}$$

Subtracting Fractions: When you subtract fractions that have the same denominators, you subtract only the numerators and keep the denominator the same.

If the fractions have diffrent denominators;

✓ Find the smallest multiple (LCD) of both numbers.

✓ Rewrite the fractions as equivalent fractions with the LCD as the denominator.

Key: $\frac{a}{b}-\frac{c}{b}=\frac{a-c}{b}$

$\frac{a}{b}-\frac{c}{d}=\frac{a\cdot(d)}{b\cdot(d)}-\frac{c\cdot(b)}{d\cdot(b)}=\frac{ad-cb}{bd}$

Multiplying Fractions:

Key: $\frac{a}{b}\cdot\frac{c}{d}=\frac{a\cdot(c)}{b\cdot(d)}=\frac{ac}{bd}$

Step 1: Multiply the numerators

Step 2: Multiply the denominators

Step 3: Simplify the fraction if needed

Example: $\frac{1}{4}\cdot\frac{1}{7}?$

Solution: $\frac{1}{4}\cdot\frac{1}{7}=\frac{1}{28}$

Dividing Fractions:

$$\frac{a}{b} \div \frac{c}{d} = \frac{a \cdot (d)}{b \cdot (c)} = \frac{ad}{bc}$$

Step 1: Flip the divisor (the second fraction)

Step 2: Multiply the first fraction by that reciprocal

Step 3: Simplify the fraction if need

Example: $\frac{1}{3} \div \frac{2}{9}$?

Solution:

Step 1: Flip the divisor $\frac{2}{9} \longrightarrow$ becomes $\frac{9}{2}$

Step 2: Multiply the first fraction by that reciprocal

$$\frac{1}{3} \cdot \frac{9}{2} = \frac{9}{6}$$

Step 3: Simplify the fraction.

$$\frac{9}{6} = \frac{3}{2} = 1\frac{1}{2}$$

Fractions Worksheet

Find the value of each expression in lowest terms.

1. $\frac{5}{10} - \frac{1}{4} =$

2. $\frac{3}{7} + \frac{5}{4} =$

3. $\frac{2}{3} + \frac{1}{11} =$

4. $\frac{5}{18} - \frac{3}{72} =$

5. $\frac{2}{6} \div \frac{1}{3} =$

6. $\frac{2}{9} \cdot \frac{1}{4} =$

7. $\frac{2}{12} \cdot \frac{2}{8} =$

8. $\frac{1}{8} \div \frac{2}{9} =$

Fractions Challenge

1. Which of the following fractions has the largest value?

A) $\frac{-1}{9}$ B) $\frac{-2}{9}$ C) $\frac{-4}{9}$ D) $\frac{-8}{9}$

2. Which group of fractions is listed from smallest to greatest?

A) $\frac{1}{9}<\frac{-1}{9}<\frac{1}{3}$ B) $\frac{-1}{9}<\frac{1}{9}<\frac{1}{3}$ C) $\frac{1}{3}<\frac{-1}{9}<\frac{1}{9}$ D) $\frac{-1}{9}<\frac{1}{3}<\frac{1}{9}$

3. Which is the equivalent multiplication problem for $2\frac{1}{3}\div 3\frac{3}{4}$?

A) $\frac{7}{3}\times\frac{4}{15}$ B) $\frac{3}{7}\times\frac{15}{4}$ C) $\frac{7}{3}\times\frac{15}{4}$ D) $\frac{7}{15}\times\frac{3}{4}$

4. The expression $\frac{4}{x+6}$ is undefined when x is equal to:

A) 4 B) −4 C) 6 D) −6

5. Find the value of the following expression in lowest terms.

$$\frac{22}{15}\div\frac{11}{5}$$

A) $\frac{1}{3}$ B) $\frac{2}{3}$ C) $\frac{3}{5}$ D) $\frac{3}{4}$

6. Find the value of the following expression in lowest terms.

$$\frac{2}{5}+\frac{4}{3}$$

A) $1\frac{1}{15}$ B) $\frac{11}{15}$ C) $11\frac{1}{15}$ D) $1\frac{11}{15}$

Fractions Challenge

7. Find the value of the following expression in lowest terms.

$$\frac{3}{4}+\frac{2}{5}$$

A) $1\frac{1}{3}$　　B) $\frac{3}{20}$　　C) $1\frac{3}{20}$　　D) $6\frac{2}{3}$

8. Find the value of the following expression in lowest terms.

$$1\frac{3}{4}\div4\frac{2}{3}$$

A) $2\frac{1}{3}$　　B) $\frac{7}{3}$　　C) $2\frac{2}{3}$　　D) $\frac{3}{8}$

9. Find the value of the following expression in lowest terms.

$$4\frac{3}{4}-3\frac{2}{5}$$

A) $1\frac{7}{20}$　　B) $1\frac{17}{20}$　　C) $2\frac{7}{20}$　　D) $\frac{7}{20}$

10. Tony and Jenny were told to find the product $4\times\frac{1}{9}$

Tony ⟶ $4\times\frac{1}{9}=\frac{1}{36}$

Jenny ⟶ $4\times\frac{1}{9}=\frac{4}{9}$

Which student wrote the product correctly?

A) Tony　　B) Jenny

Fractions Challenge

11. Find the product of $\frac{5}{6}$ and $\frac{6}{7}$

A) $\frac{35}{36}$ B) $\frac{5}{7}$ C) $\frac{7}{5}$ D) $\frac{3}{5}$

12. A grocery store bought 45 pounds of tomatoes and sold $\frac{5}{9}$ on the same day. At the end of the day, how many pounds of tomatoes where left?

A) 25 pounds B) 20 pounds C) 45 pounds D) 81 pounds

13. The road is 48 miles long, and $\frac{1}{3}$ of the road is paved. Another $\frac{1}{4}$ of the road is paved. How much of the road is left unfinished?

A) 12 miles B) 24 miles C) 32 miles D) 36 miles

14. There are 24 students in math class. If $\frac{2}{3}$ of the students in this class are male students, find the number of female students in the math class?

A) 6 B) 8 C) 12 D) 16

15. Which of following is $\frac{2}{3}$?

A) 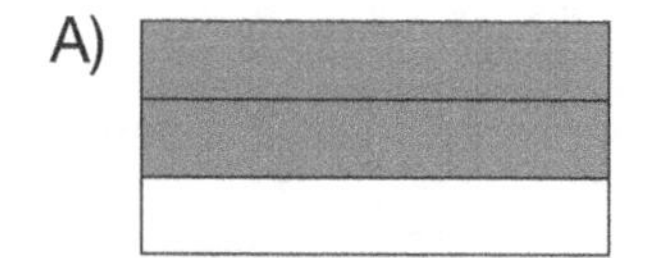 B) 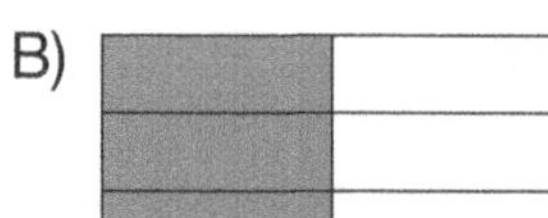 C) 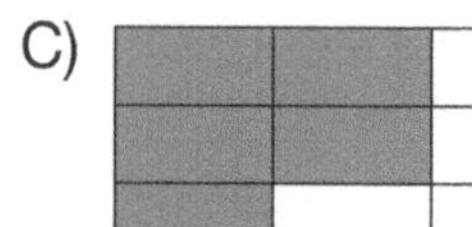 D) 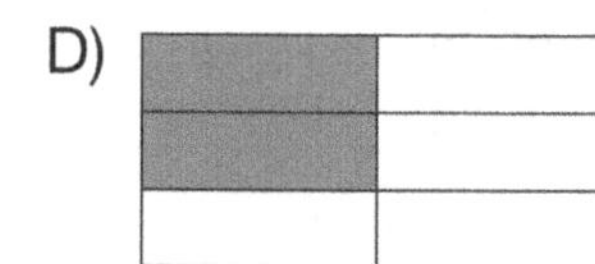

16. The width of a rectangular garden is $1\frac{2}{3}$ cm. The length is $1\frac{4}{5}$ cm. Which of the following is the area of the garden?

A) $\frac{1}{3}$ cm^2 B) 3 cm^2 C) $\frac{2}{3}$ cm^2 D) 4 cm^2

17. Find the product of $1\frac{1}{2} \times 3\frac{1}{5}$.

A) $4\frac{4}{5}$ B) $5\frac{1}{5}$ C) $4\frac{1}{5}$ D) $3\frac{1}{10}$

18. Tony has 12 pencils and wants to give $\frac{3}{4}$ of them to a friend while keeping the rest for himself. How many pencils would his friend get?

A) 3 B) 6 C) 8 D) 9

19. Last night, Tony spent $1\frac{1}{8}$ hours doing his math homework. John did his math homework for $\frac{1}{4}$ as many hours as Tony did. How many hours did John spent on his homework?

A) $\frac{9}{32}$ B) $4\frac{1}{2}$ C) $\frac{3}{8}$ D) $\frac{3}{16}$

20. Find the quotient $\frac{1}{7} \div 3$.

A) $\frac{3}{7}$ B) $2\frac{1}{3}$ C) 21 D) $\frac{1}{21}$

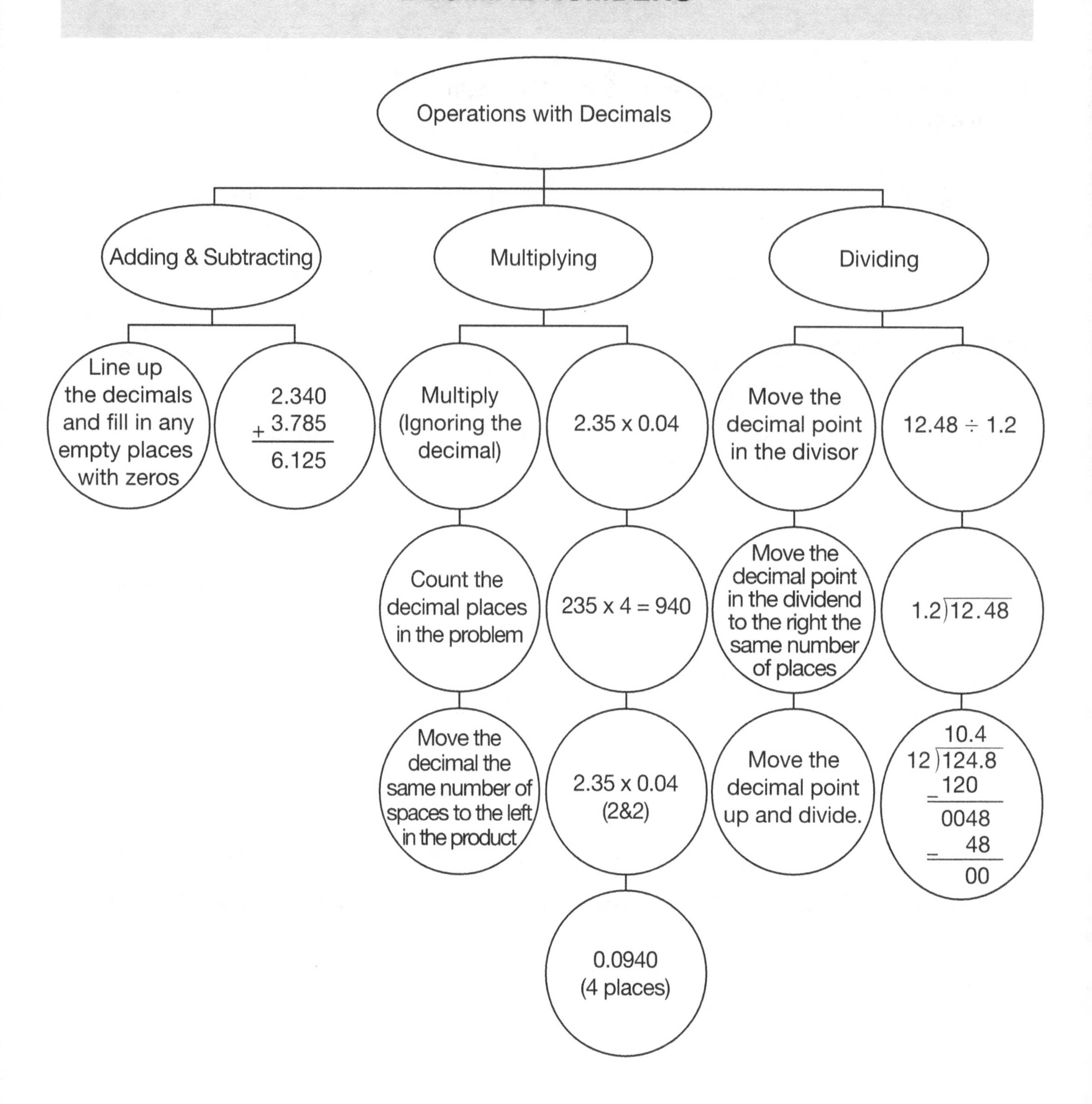

Operations with Decimals
Adding & Subtracting
Multiplying
Dividing
Line up the decimals and fill in any empty places with zeros
2.340
+ 3.785
6.125
Multiply (Ignoring the decimal)
2.35 x 0.04
Move the decimal point in the divisor
12.48 ÷ 1.2
Count the decimal places in the problem
235 x 4 = 940
Move the decimal point in the dividend to the right the same number of places
1.2)12.48
Move the decimal the same number of spaces to the left in the product
2.35 x 0.04
(2&2)
Move the decimal point up and divide.
10.4
12)124.8
– 120
0048
– 48
00
0.0940
(4 places)

Decimal Numbers Worksheet

1. Add each decimal to the given number.	**2.** Subtract each decimal from the given number.
$3.5 + 4.4 =$	$8.9 - 4.8 =$
$13.4 + 5.7 =$	$18.90 - 3.9 =$
$6.85 + 6.78 =$	$72.8 - 6.86 =$
$345.56 + 56.8 =$	$47.45 - 22.65 =$
$43.70 + 121.567 =$	$101.567 - 89.568 =$
3. Multiply each decimal with the given number.	**4.** Divide each decimal by the given number.
$3.8 \times 7 =$	$4.8 \div 4 =$
$13.8 \times 5 =$	$72.6 \div 3 =$
$3.8 \times 20 =$	$124.8 \div 4 =$
$4.8 \times 7.2 =$	$36.8 \div 0.8 =$
$12.7 \times 6.74 =$	$24.8 \div 0.04 =$

Decimal Numbers Challenge

Find each sum or difference. Show your work.

1. $12.5 + 13.9 =$

2. $24.54 + 34.73 =$

3. $78.56 + 18.458 =$

4. $8.6 - 3.74 =$

5. $18.48 - 9.6 =$

6. $33.54 - 4.6789 =$

Find each product or quotient. Show your work.

7.
$$\begin{array}{r} 3.58 \\ \times\ 4.3 \\ \hline \end{array}$$

8. $0.8\overline{)12.8}$

9.
$$\begin{array}{r} 18.50 \\ \times\ 4.03 \\ \hline \end{array}$$

10. $3.6\overline{)18.18}$

11. Elif and her 4 friends ate lunch at a restaurant. They decided to split the bill equally. The total bill was $24.85. How much was each person's share?

12. Vera paid $9 for a bag of potatoes. The potatoes cost $.45 per lb. How much did the bag of potatoes weigh?

ROUNDING NUMBERS

✓ Identify the units digit

✓ Round up or down

✓ If the digit is 5 or greater, add one more.

✓ If the digit is less than 5, leave it the same.

Example: Round 68 to the nearest tens.

Solution:

✓ Keep the 6.

✓ The next digit is "8" which is 5 or more, so increase the "6" by 1 to 7

✓ The answer is 70.

Place Value Chart										
Millions	Hundred Thousands	Ten Thousands	Thousands	Hundreds	Tens	Ones	Decimal Point	Tenths	Hundredths	Thousandths
1	3	5	6	7	8	9	•	7	6	4
Whole Number								Decimal Number		
One million, three hundred fifty six thousand, seven hundred eighty nine and 7 hundred sixty four thousandths										

Rounding Numbers Worksheet

Round each number to the place value of the underlined digit.

1. 6,780,089

2. 34,678,457

3. 503,132

4. 7,184

5. 2,345,678

6. 375

Round each number to the place value indicated.

7. 456,346; hundreds

8. 1,234; tens

9. 3,547,892; thousands

10. 12,345,756; millions

Rounding Numbers Challenge

1. Round the whole number to the given place.
48 to the nearest ten

A) 47 B) 48 C) 49 D) 50

2. Round the whole number to the given place.
5,368 to the nearest thousand

A) 4000 B) 5000 C) 6000 D) 6500

3. Round the perimeter of the following triangle to the nearest ten.

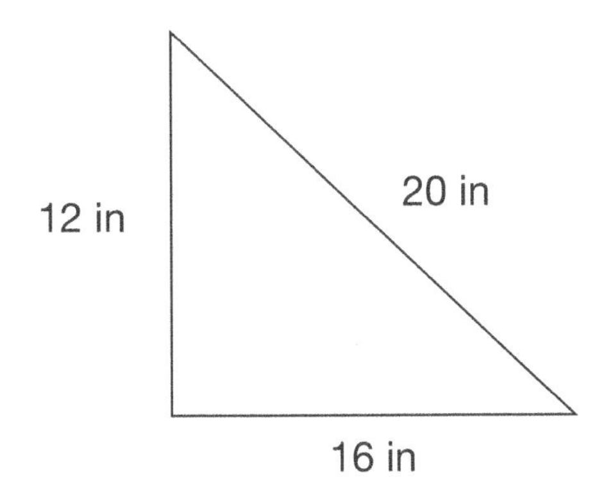

A) 45 in B) 50 in C) 58 in D) 60 in

For question 4-8, use the table.

Population of Colleges in 2020 Census	
Colleges	**Population**
College A	26,458
College B	26,690
College C	26,049

4. Which college has a population that rounds to 26,000?

5. What is the population of college A rounded to the nearest thousand?

6. What is the population of college B rounded to the nearest hundred?

7. What is the population of college C rounded to the nearest ten?

8. Which college has the highest population?

9. Which number rounds to 25,000?

A) 24,399 B) 24,098 C) 24,890 D) 25,987

10. A number rounded to the nearest hundred is 18,600. Determine the largest possible number.

A) 18,000 B) 18,584 C) 18,648 D) 18,649

LAWS OF EXPONENTS

1. **Product Rule:** When multiplying exponents with the same base, always add the powers and keep the base the same.

$a^x \cdot a^y = a^{x+y}$

Examples:

$3^5 \cdot 3^8 = 3^{5+8} = 3^{13}$

$3^{-5} \cdot 3^{-8} = 3^{-5+(-8)} = 3^{-13}$

$4^{-5} \cdot 4^8 = 4^{8-5} = 4^3$

2. **Quotient Rule:** When dividing exponents with the same base, always subtract the powers and keep the base the same.

$a^x \div a^y = a^{x-y}$

Examples:

$5^5 \div 5^3 = 5^{5-3} = 5^2$

$7^5 \div 7^{-3} = 7^{5-(-3)} = 7^8$

$5^5 \div 5^3 = 5^{5-3} = 5^2$

$11^{-5} \div 11^{-3} = 11^{-5-(-3)} = 11^{-5+3} = 11^{-2} = \frac{1}{11^2} = \frac{1}{121}$

3. **Power to Power Rule:** When you have one base but more than one power, keep the base same and multiply all powers together.

$(a^x)^y = a^{xy}$

Example:

$(3^5)^7 = 3^{5\cdot 7} = 3^{35}$

4. **Negative Exponent:** When you have a negative power, always take the reciprocal of the base.

$a^{-x} = \frac{1}{a^x}$

Examples:

$3^{-4} = \frac{1}{3^4} = \frac{1}{81}$

$-3^{-2} = \frac{1}{(-3)^2} = \frac{1}{9}$

5. **Zero Exponent:** Any number to the zero power is always 1.

$a^0 = 1$

Examples:

$5^0 = 1$ $(-12)^0 = 1$ $\left(\frac{1}{7}\right)^0 = 1$

Laws of Exponents Worksheet

Identify the base and exponent in each of the following

1. 2^4

Base: ______________

Exponent: ______________

2. a^{3b}

Base: ______________

Exponent: ______________

3. $\left(\frac{1}{2}\right)^3$

Base: ______________

Exponent: ______________

Write the numerals in exponential form with the given base and exponent.

4.

Base: 3

Exponent: 5

Exponential form: __________

5.

Base: 6

Exponent: −2

Exponential form: __________

6.

Base: a

Exponent: 4b

Exponential form: __________

Express each of the following in standard form.

7. $3^4 =$ ______

8. $\left(\frac{1}{6}\right)^{-3} =$ ______

9. $\left(\frac{3}{2}\right)^{-4}$

Laws of Exponents Challenge

1. If $3^x \cdot 3^x \cdot 3^x = 27^4$, then find x.

A) 1 B) 2 C) 3 D) 4

2. Which of the following is equal to $\frac{8^2 \cdot 16^3}{2^{10}}$?

A) 2^8 B) 2^7 C) 2^5 D) 2^4

3. $(-1)^2 \cdot (-1)^3 \cdot (-1)^{25}$ is equivalent to:

A) 1 B) -1 C) 3 D) 4

4. Which of the following is equivalent to $\left(\frac{x+4}{3}\right)^0$, when $x \neq -4$?

A) 0 B) 1 C) 2 D) 3

5. $\left(\left(-\frac{2}{3}\right)^4\right)^5$ is equivalent to:

A) $\left(\frac{3}{2}\right)^{20}$ B) $\left(\frac{2}{3}\right)^{20}$ C) $\left(\frac{4}{3}\right)^{20}$ D) 3^{20}

Laws of Exponents Challenge

6. Evaluate $3^{12} \cdot 3^{7} = 3^{n} \cdot 3^{4} \cdot 3^{m}$, then find $m + n$

A) 10 B) 12 C) 14 D) 15

7. Evaluate $2^{12} \cdot 2^{7} \cdot 2^{4} \cdot 2^{8}$

A) 2^{13} B) 2^{16} C) 2^{21} D) 2^{31}

8. Evaluate $(-4)^{2} \cdot (-2)^{6}$

A) 2^{3} B) 2^{6} C) 2^{8} D) 2^{10}

9. Evaluate $\left(\frac{3}{2}\right)^{-4}$

A) $\frac{81}{16}$ B) $\frac{16}{81}$ C) $-\frac{81}{16}$ D) $-\frac{16}{81}$

10. Evaluate $(x^{4} v^{5}) \cdot (x^{-3} v^{-6})$

A) $\frac{x}{v}$ B) $\frac{v}{x}$ C) $-\frac{x}{v}$ D) $-\frac{v}{x}$

LAWS OF RADICALS

1. $(\sqrt[n]{a})^n = a$

Example: $(\sqrt[3]{4})^3 = 4$

2. $\sqrt[n]{ab} = \sqrt[n]{a} \cdot \sqrt[n]{b}$

Example: $\sqrt[2]{3 \cdot 4} = \sqrt[2]{3} \cdot \sqrt[2]{4}$

3. $\sqrt[n]{a} \cdot \sqrt[n]{b} = \sqrt[n]{a \cdot b}$

Example: $\sqrt[2]{7} \cdot \sqrt[2]{5} = \sqrt[2]{7 \cdot 5} = \sqrt{35}$

4. $(\sqrt[n]{a^m})^p = \sqrt[n]{a^{mp}}$

Example: $(\sqrt[5]{3^4})^2 = \sqrt[5]{3^{4 \cdot 2}} = \sqrt[5]{3^8}$

5. $\sqrt[n]{\frac{a}{b}} = \frac{\sqrt[n]{a}}{\sqrt[n]{b}}$, $b \neq 0$

Example: $\sqrt[4]{\frac{3}{5}} = \frac{\sqrt[4]{3}}{\sqrt[4]{5}}$

6. $\sqrt[m]{\sqrt[n]{a}} = \sqrt[m \cdot n]{a}$

Example: $\sqrt[2]{\sqrt[3]{5}} = \sqrt[2 \cdot 5]{5} = \sqrt[10]{5}$

7. $\frac{1}{\sqrt[n]{a}} = \frac{\sqrt[n]{a^{n-1}}}{a}$, $a \neq 0$

Example: $\frac{1}{\sqrt[2]{3}} = \frac{\sqrt[2]{3^{2-1}}}{3} = \frac{\sqrt[2]{3}}{3} = \frac{\sqrt{3}}{3}$

Laws of Radicals Worksheet

Simplify each radical expression. Show your work.

1. $\sqrt{\frac{36}{64}}$ = ____________

2. $\sqrt{\frac{6}{3}}$ = ____________

3. $\sqrt{\frac{3}{5}}$ = ____________

4. $\frac{2}{\sqrt{5}}$ = ____________

5. $\frac{\sqrt{6}}{\sqrt{8}}$ = ____________

6. $\sqrt{\frac{24}{3}}$ = ____________

7. $2\sqrt{20}+\sqrt{5}$ = ____________

8. $2\sqrt{18}-\sqrt{12}$ = ____________

9. $\sqrt{121}+\sqrt{144}+\sqrt{169}$ = ____________

10. $-3\sqrt{2}+\sqrt{2}+\sqrt{8}$ = ____________

11. $\sqrt{121}+\sqrt{225}+\sqrt{400}$ = ____________

12. $-3\sqrt{3}+\sqrt{3}+\sqrt{27}$ = ____________

13. $3\sqrt{45}-9\sqrt{5}+\sqrt{75}$ = ____________

14. $-3\sqrt{3}+\sqrt{27}+\sqrt{12}$ = ____________

Laws of Radicals Challenge

1. Simplify $8\sqrt{6}-3\sqrt{24}=?$

A) $\sqrt{6}$ B) $2\sqrt{6}$ C) $3\sqrt{6}$ D) $4\sqrt{6}$

2. Simplify $3\sqrt{5}\cdot\sqrt{15}=?$

A) $15\sqrt{3}$ B) $3\sqrt{15}$ C) $5\sqrt{3}$ D) $3\sqrt{5}$

3. Simplify $\frac{7\sqrt{12}}{\sqrt{3}}$

A) 7 B) 14 C) $7\sqrt{2}$ D) 21

4. If $x=\frac{2^5}{\sqrt{8}}$, then find x.

A) $2\sqrt{2}$ B) $4\sqrt{2}$ C) $6\sqrt{2}$ D) $8\sqrt{2}$

5. What is the solution of the equation given below?

$$\frac{\sqrt{2}\cdot 6^{\frac{1}{3}}}{\sqrt[3]{3}\cdot 2^{-\frac{1}{6}}}$$

A) 1 B) 2 C) 3 D) 4

6. If $x > 0$ and $x - 3 = \sqrt{x-3}$, then which of the following can be x?

A) 0 B) 1 C) 2 D) 4

7. If $a = \sqrt{3}$ and $b = \sqrt{2}$, then find $\frac{a}{b} - \frac{b}{a}$.

A) $\frac{\sqrt{6}}{6}$ B) $\frac{1}{\sqrt{5}}$ C) $\frac{1}{\sqrt{3}}$ D) $\frac{1}{\sqrt{2}}$

8. What is the solution(s) for y in the following equation?

$$\frac{12}{\sqrt[5]{y}} = 6$$

A) 16 and −16 B) 32 and −32 C) 16 only D) 32 only

9. Simplify $\frac{\sqrt{a}}{2\sqrt{a}}$

A) $\frac{2\sqrt{a}+a}{4-a}$ B) $\frac{\sqrt{a}}{4}$ C) $\frac{a}{2}$ D) $\frac{1}{2}$

10. Simplify $\sqrt{2}(\sqrt{12} - \sqrt{3})$

A) $\sqrt{2}$ B) $\sqrt{3}$ C) $\sqrt{6}$ D) $2\sqrt{3}$

SCIENTIFIC NOTATION

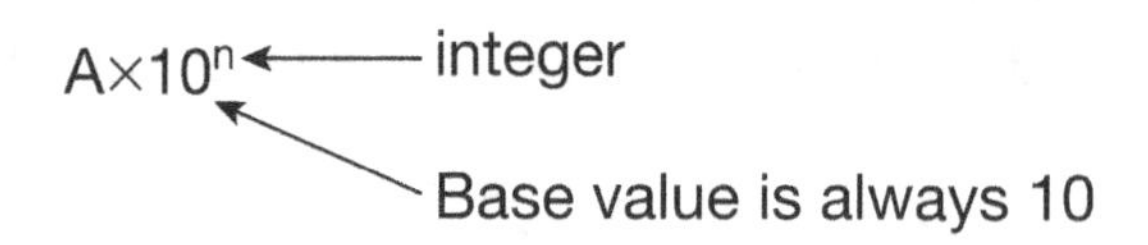

$1 \leq |A| < 10$

A is a number greater than or equal to 1 but less than 10

Examples:

1. $0.00015 = 1.5 \times 10^{-4}$

2. $156{,}000 = 1.56 \times 10^5$

3. $685 \times 10^{-3} = 6.85 \times 10^{-1}$

4. $123{,}000{,}000 = 1.23 \times 10^8$

5. $12 = 1.2 \times 10^1$

6. $5 = 5 \times 10^0$

Scientific Notation Worksheet

Write each exponent in scientific notation.

1. $125 \times 10^{5} =$ __________

2. $0.0123 \times 10^{5} =$ __________

3. $98 \times 10^{-6} =$ __________

4. $145{,}000 \times 10^{3} =$ __________

5. $0.0457 \times 10^{-9} =$ __________

6. $86.9 \times 10^{-12} =$ __________

Write each exponent in scientific notation.

7. $967 \times 10^{4} =$ __________

8. $0.0457 \times 10^{-1} =$ __________

9. $21.9 \times 10^{3} =$ __________

10. $1.5 \times 10^{0} =$ __________

11. $0.04 \times 10^{6} =$ __________

12. $3.45 \times 10^{7} =$ __________

Write each number in scientific notation.

13. $1.8 =$ __________

14. $0.0444 =$ __________

15. $3.784 =$ __________

16. $89{,}000{,}000 =$ __________

17. $3.67 =$ __________

18. $234{,}000 =$ __________

Scientific Notation Challenge

1. What is 0.00025 written in scientific notation form?

A) 2.5 B) 25×10^5 C) 25×10^4 D) 2.5×10^{-4}

2. Which of the following is a number written in scientific notation form?

A) 10×10^4 B) 25×10^6 C) 0.5×10^{-4} D) 2.5×10^{-4}

3. What is 183×10^{-3} written in standard form?

A) 183 B) 183.3 C) 1.83 D) 0.183

4. The length of the Mississippi River is 3,730 kilometers. What is the length written in scientific notation?

A) 0.373×10^4 B) 3.73×10^3 C) 37.3×10^2 D) 373×10^1

5. The population of Istanbul is approximately 15×10^6 people. The population of Turkey is approximately 81×10^6 people. About how many times greater is the population of the Turkey population of Istanbul? (Write the answer in scientific notation form)

A) 54×10^1 B) 5.4×10^0 C) 5.4×10^1 D) 0.54×10^1

6. For which value of n is the equation below true?
$5.4 \times 10^{25} = 54 \times 10^{3n}$

A) 4 B) 6 C) 8 D) 9

7. Multiply $(2.4 \times 10^{-5}) \times (5 \times 10^{12})$ and write the answer in scientific notation.

A) 12×10^{7} B) 1.2×10^{8} C) 1.2×10^{7} D) 0.12×10^{9}

8. Divide $(18.6 \times 10^{14}) \div (3 \times 10^{12})$ and write the answer in scientific notation.

A) 6.2×10^{2} B) 6.2×10^{3} C) 63×10^{1} D) 0.63×10^{4}

9. What is 1.83 written in scientific notation?

A) 1.83×10^{3} B) 1.83×10^{-1} C) 1.83×10^{1} D) 1.83×10^{0}

ALGEBRAIC EXPRESSIONS

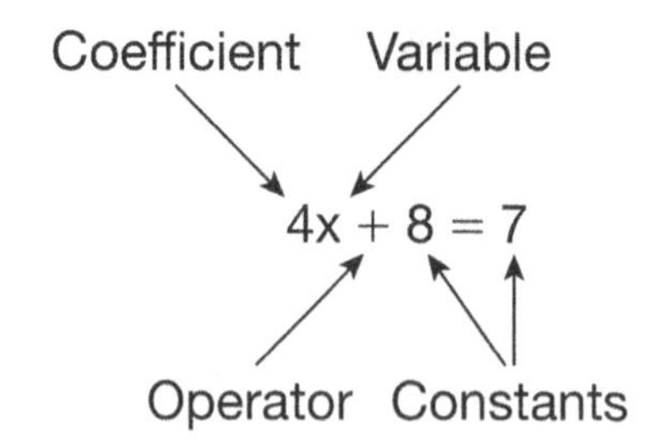

Variable: A letter that represents an unknown number. Any letter can be used as a variable.

Example: x, y, z...etc.

Coefficient: A number placed before a variable.

Example: $2x$, $3(x + 7)$

Constant: Any real number used in an expression or equation.

Example: $6x + 7$

Terms: Terms are separated by a positive or negative sign in an expression.

Example: $x + y + 3$ has 3 terms: x, y, and 3.

Note: x and y are variable terms and 3 is a constant term.

Like terms: Like terms are terms that have the same variables with the same exponents.

Example: $6x$ and $5x$

Algebraic Expressions Worksheet

Simplify following each of the following expressions.

1. $4x - 3x =$ ______________

2. $7x - x =$ ______________

3. $-9x + 4x =$ ______________

4. $1 + 3n - 4n =$ ______________

5. $7 - 3n + 10 =$ ______________

6. $-3n - 6n =$ ______________

7. $2(3 - 4v) =$ ______________

8. $-2(-3v + 6) =$ ______________

9. $-3 - 4(v - 7) =$ ______________

10. $2d - 4 - 3d - 8 =$ ______________

11. $8(3d + 6) - 12 =$ ______________

12. $\frac{1}{2}(d - 12) =$ ______________

13. $-\frac{3}{2}(x - 8) =$ ______________

14. $\frac{1}{5}(15x - 25) =$ ______________

15. $\frac{x}{2}(x - 2) =$ ______________

16. $\frac{1}{3}(12x + 9) =$ ______________

Algebraic Expressions Challenge

1. Find the perimeter of the trapezoid below.

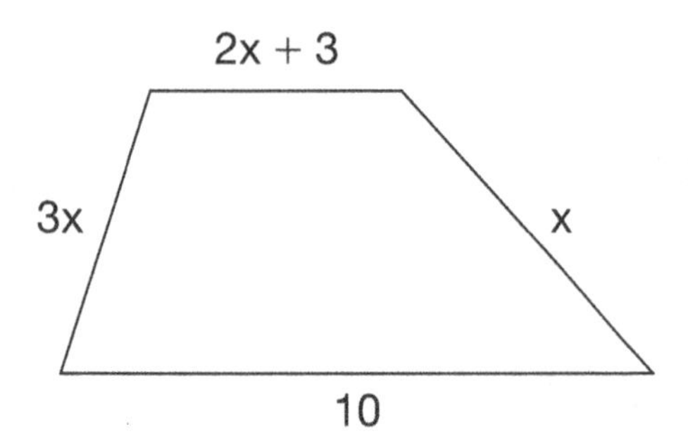

A) $6x + 13$ B) $6x + 3$ C) $6x + 10$ D) $3x + 13$

2. Simplify $9x + 11y - 6x + 7y = ?$

A) $3x + 18y$ B) $18x + 3y$ C) $18x - 3y$ D) $3x - 18y$

3. Simplify $-3(4x - 6) -5(x + 1) = ?$

A) $17x - 13$ B) $-17x + 13$ C) $12x + 13$ D) $7x - 19$

4. Write an expression for the area of the following figure.

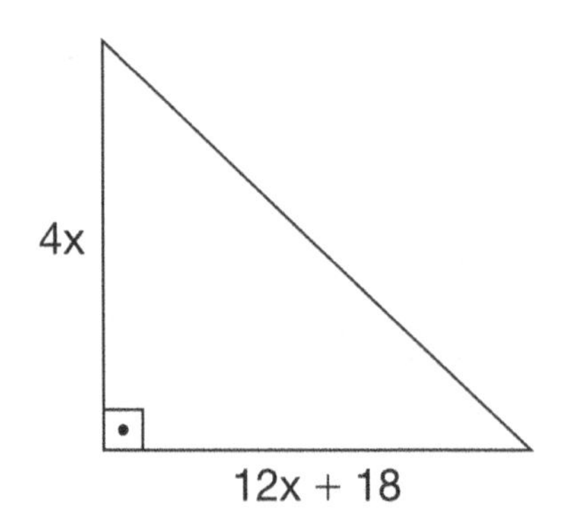

A) $24x^2 - 36x$ B) $24x^2 + 36x$ C) $48x^2 - 72$ D) $48x^2 + 72x$

Algebraic Expressions Challenge

5. What is the cost of 4 notebooks at x dollars and 7 pencils at y dollars each?

A) $4x - 7y$ B) $7x$ C) $7y$ D) $4x + 7y$

6. Write the following algebraic expression in its simplest form.

$\frac{3x}{2}(8y - 18)$

A) $12xy - 9x$ B) $12xy - 27x$ C) $12xy + 27x$ D) $12xy - 18x$

7. Write the following algebraic expression in its simplest form.

$12x + 2y + 8x + 4y$

A) $20x + 6y$ B) $20x + 4y$ C) $26xy$ D) $14x + 12y$

8. Simplify $\frac{28x^2y}{14xy}$

A) $2xy$ B) $2x$ C) $2y$ D) $-2xy$

9. Simplify $\frac{10x}{y} + \frac{5x}{2y}$

A) $25xy$ B) $\frac{25x}{y}$ C) $\frac{25x}{2y}$ D) $-25xy$

10. Simplify $\frac{ab}{b} \div \frac{a}{2b}$

A) a B) b C) $2a$ D) $2b$

EQUATIONS WITH TWO VARIABLES

Linear Equations: A set of two or more linear equations containing two or more variables.

Variable: Represents an unknown number. Any letter can be used as a variable.

Example: x, y, z, ... etc.

Coefficient: A number placed before a Variable in an equation.

Example: (3)x, (5)(x + 5)

Constant: A symbol which represents a specific number.

Example: (5), 3x + (10)

Equation: A sentence that states that two mathematical expressions are equal.

Example: $4x - 6 = 10$

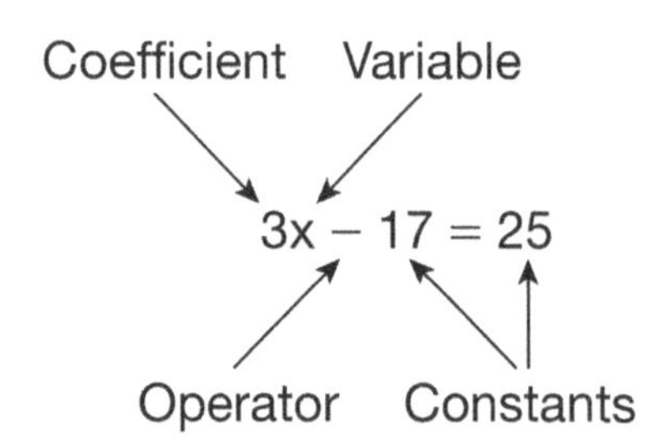

Equations with Two Variables Worksheet

Determine the value of the unknown variable in each equation.

1. $15 - x = y$

$x = 4, y =$ ______

2. $6 + 2x = y$

$x = 8, y =$ ______

3. $2x - 4 = y$

$y = 10, x =$ ______

4. $11x = y$

$y = 55, x =$ ______

5. $2x = y - 10$

$x = 8, y =$ ______

6. $x = y - 12$

$y = 10, x =$ ______

7. $\frac{1}{x} = y$

$y = 18, x =$ ______

8. $\frac{3}{4}x = y$

$x = 8, y =$ ______

9. $x = -\frac{5}{3}y - 12$

$y = 9, x =$ ______

10. $5x = -3y$

$y = 15, x =$ ______

11. $\frac{x}{2} = y - 10$

$x = 12, y =$ ______

12. $2x - y = 12$

$y = 14, x =$ ______

Equations with Two Variables Challenge

1. If $2x + 3y = 18$, find x when $y = 4$.

A) 2 B) 3 C) 6 D) 8

2. If $\frac{x}{y-2} = 10$, find x when $y = 6$.

A) 20 B) 30 C) 40 D) 50

3. If $x + \sqrt{y} = 16$, find x when $y = 25$.

A) -9 B) 9 C) 11 D) 21

4. If $x^2 - y = 45$, find y when $x = 9$.

A) 26 B) 28 C) -36 D) 36

5. If $x + 3(y - 4) = 24$, find x when $y = 2$.

A) 24 B) 28 C) -30 D) 30

6. If $\frac{x-5}{y} = 4$, find x when $y = -2$.

A) -3 B) 3 C) -8 D) 8

7. If x is a postive integer and $x^2 = 4y$, find x when $y = 9$.

A) 6 B) -6 C) 18 D) 36

8. If $x = \frac{2y-4}{10}$, find x when $y = 10$.

A) 1.2 B) 1.6 C) 1.8 D) 2.2

9. Jenny is 4 years younger than twice her sister's age. If Jenny is 30 years old, then how old is Jenny's sister?

A) 15 B) 16 C) 17 D) 18

Solving Equations

Equation: A sentence in which two mathematical expressions are equal.

Tips for solving equations:

✓ Add or subtract to combine like terms.

✓ Multiply or divide to solve for the variable.

✓ Check your solution.

Example:

$2x + 7 = 13$

$\quad + -7 \quad -7$ ⟶ (subtract −7 on both sides)

$2x = 6$

$\frac{2x}{2} = \frac{6}{2}$ ⟶ (divide by 2 on both sides)

$x = 3$

$2(3) + 7 = 13$ ⟶ check

$13 = 13$

Inequalities

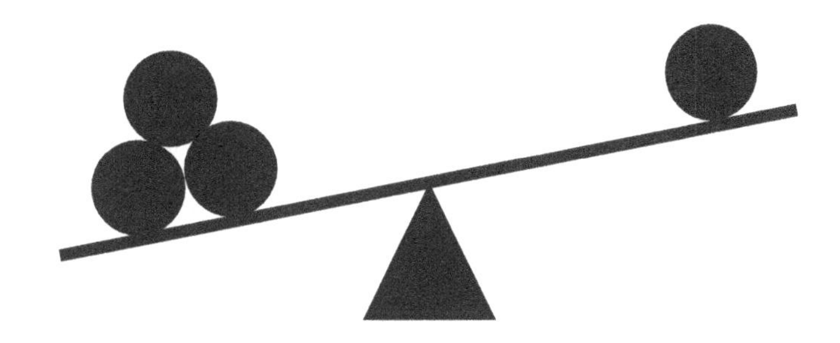

Inequality Symbols	Number Line Symbols
$\leq$	←——●
$\geq$	●——→
$<$	←——○
$>$	○——→

Study Tips:

1. If $a < x$ and $x < b$, then $a < x < b$
2. If $a \leq b$, then $a < b$ or $a = b$
3. If $a \geq b$, then $a > b$ or $a = b$
4. If $a < b$ and $c < d$, then $a + c < b + d$
5. If $\frac{1}{a} < \frac{1}{x} < \frac{1}{b}$, then $a > x > b$ (a, x and b are positive integers)

Solving Equations & Inequalities Worksheet

Solve each one–step equation.

1. $12 = 3.2x$

$x =$ ______

2. $-18 = 9x$

$x =$ ______

3. $\frac{x}{-3} = -6$

$x =$ ______

Solve each two–step equation.

4. $2x + 1 = 13$

$x =$ ______

5. $14 - 3x = 26$

$x =$ ______

6. $\frac{x-5}{-2} = 8$

$x =$ ______

Solve each multi–step equation.

7. $\frac{1}{2}x + 2x - 3 = 7$

$x =$ ______

8. $12x - 3(x - 5) = 42$

$x =$ ______

9. $\frac{x-4}{-2} - \frac{3x}{2} = 4$

$x =$ ______

Solve and graph each one–step inequality.

10. $x - 4 < 8$

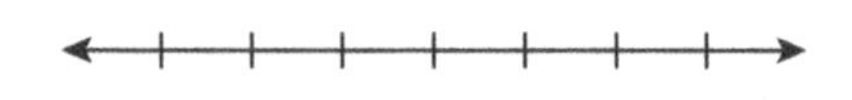

11. $4x < -12$

12. $10x - 15(2x + 1) > 25$

Solving Equations & Inequalities Challenge

1. What is the value of k in the equation shown below?

$3k - 5 + 4k = -4(k + 4)$

A) 0 B) 1 C) −1 D) 2

2. What is the value of x in the equation shown below?

$\frac{1}{3}(x - 12) + 6 = 2x - 8$

A) 0 B) 1 C) 2 D) 6

3. $(2x + 4) - (4x - 3) = 19$

What is the value of x in the equation shown above?

A) −8 B) −6 C) 6 D) 5

4. $7x - 8 \geq 3(x - 6)$

Which of the following numbers is **NOT** a solution of the above inequality?

A) −3 B) −2 C) −1 D) 0

5. Solve the following inequality.

$3x - (x + 5) > -2(2 + x) + 7$

A) $x < 2$ B) $x > 2$ C) $x < -2$ D) $x > -2$

6. What is the solution to the equation below?

$-3(2y - 5) = 6(3y - 12 \cdot 5)$

A) 3 B) 4 C) $\frac{4}{15}$ D) $\frac{15}{4}$

Solving Equations & Inequalities Challenge

7. For the following inequality, which of the following **MUST** be true?

$\frac{x}{5} - 2 < 1$

A) 10 B) 15 C) 20 D) 25

8. $x < 2$

$y \geq -2$

Which of following ordered pairs (x,y) satisfies the system of equations above?

A) (−3, 2) B) (−2, −3) C) (2, −2) D) (3, −2)

9. If $6x^2 = 24$, then x can be which of following?

A) 2 B) −4 C) 4 D) 6

10. $\frac{2x+4}{3} - \frac{x}{2} = \frac{3}{2}$ Find the value of x.

A) 1 B) 3 C) 6 D) 9

11. $x = \frac{2^3}{\sqrt{64}}$ Find x.

A) 1 B) −1 C) 2 D) 4

12. What value of x satisfies the equation

$3 + \frac{x^3}{4} = 5$?

A) 2 B) 3 C) 4 D) 5

RATIOS, PROPORTIONAL RELATIONS & VARIATIONS

Ratio: A ratio is a comparison or relation between two quantities.

Note: The ratio of a and b is written as a to b, or $\frac{a}{b}$.

Example: Find the ratio of 3.5 to 4.5

Solution: ratio of 3.5 to 4.5 $= \frac{3.5}{4.5} = \frac{35}{45} \div \frac{5}{5} = \frac{7}{9}$ or 7 to 9

Proportions: A proportion is an equation that shows two equivalent ratios. If $\frac{a}{b} = \frac{c}{d}$, then, by cross multiplication, $ad = bc$

Direct variation: A direct variation is a relationship between two quantities of x and y that can be written in the following form:

$y = kx$, and $k \neq 0$ Where k is a constant of variation and k does not equal zero.

Inverse variation: An inverse variation is a relationship between two quantities of x and y that can be written in the following form:

$xy = k$, and $k \neq 0$ Where k is a constant of variation and k does not equal zero.

Ratios, Proportional Relations & Variations Worksheet

Determine whether each of the following is an equivalent ratio.

1. $\frac{1}{2}$ and $\frac{9}{18}$

2. $\frac{3}{5}$ and $\frac{27}{45}$

3. $\frac{1}{4}$ and $\frac{16}{64}$

Determine whether each question represents a direct, or inverse variation?

4. $y = 4x$

5. $xy = 13$

6. $\frac{y}{x} = 19$

Solve for x.

7. $\frac{25}{x} = \frac{35}{14}$

8. $\frac{x}{7} = \frac{18}{28}$

9. $-\frac{x}{12} = -\frac{3}{4}$

Solve for y.

10. $\frac{y+3}{4} = \frac{70}{14}$

11. $\frac{y-5}{10} = \frac{1}{2}$

12. $\frac{3y-4}{12} = \frac{1}{6}$

13. The ratio of girls to boys in art class is 4:5. If there are 32 girls, how many boys are there?

Ratios, Proportional Relations & Variations Challenge

1. There are 28 students in Mr. John's math class. They are completing their classwork and then turn to their group to discuss their work. The ratio of complete work to incomplete work was 3 to 4. How many students did not complete their classwork?

A) 3 B) 7 C) 8 D) 16

2. Find the ratio of 3.5 to 4.5

A) 9 to 7 B) 7 to 9 C) 3 to 7 D) 2 to 3

3. What is the value of $\frac{m}{n}$ if $\frac{18}{m} = \frac{12}{n}$

A) $\frac{3}{2}$ B) $\frac{2}{3}$ C) 2 D) 3

4. Which of the following graphs shows direct variation?

A) 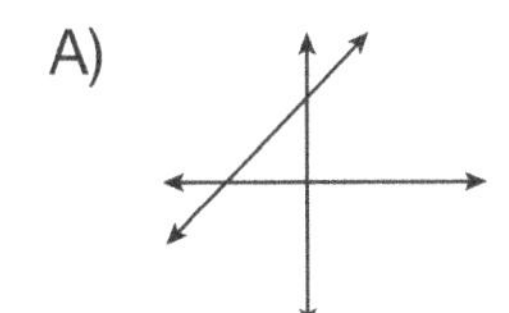

B) 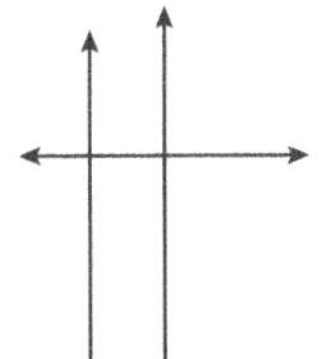

C) 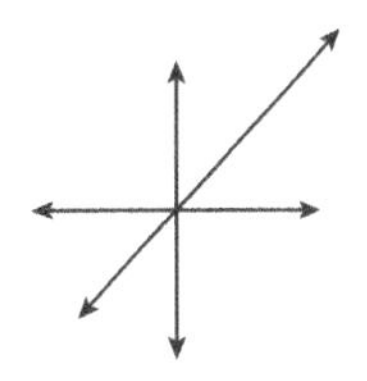

D) 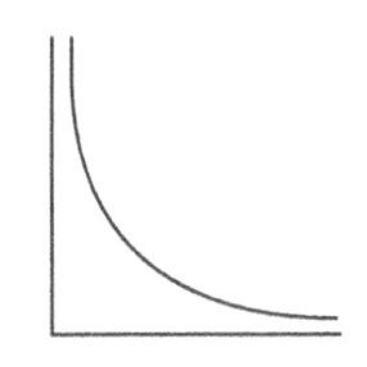

5. Find y when x = 3, if y varies directly as x, and y = 20 when x = 5.

A) 6 B) 9 C) 12 D) 15

6. If y varies inversely as x, and x = 12 when y = 60, find y when x = 18.

A) 20 B) 30 C) 40 D) 50

Ratios, Proportional Relations & Variations Challenge

7. Two numbers have a ratio of 7 to 2. If they are positive and differ by 45, what is the value of the smaller number?

A) 18 B) 21 C) 27 D) 33

8. There are 28 students in art class and they complete their classwork and then turn to the group to discuss their work. The ratio of complete work to incomplete work was 2 to 5. How many students completed their classwork?

A) 8 B) 12 C) 16 D) 20

9. If the ratio of x to y is $\frac{3}{4}$, what is the ratio of $\frac{7x}{5y}$?

A) $\frac{3}{4}$ B) $\frac{7}{5}$ C) $\frac{20}{21}$ D) $\frac{21}{20}$

10. Find the ratio of 1.5 to 3.5

A) 7 to 3 B) 3 to 7 C) 3 to 5 D) 7 to 3

11. If $\frac{x}{y} = \frac{2}{5}$, then what is $\frac{7x}{4y}$?

A) $\frac{7}{10}$ B) $\frac{5}{7}$ C) $\frac{3}{10}$ D) $\frac{10}{7}$

12. What is the solution of the following equation?

$$\frac{3}{x+1} = \frac{1}{x-1}$$

A) 2 B) 3 C) 4 D) 5

FUNCTIONS

Function: If every element of A is paired with the elements of B at least once and $A \neq \emptyset$ and $B \neq \emptyset$, this correlation is called a function.

Function Notation

Name of Function

Output ↑ →Input

$y = F(x)$

Example:

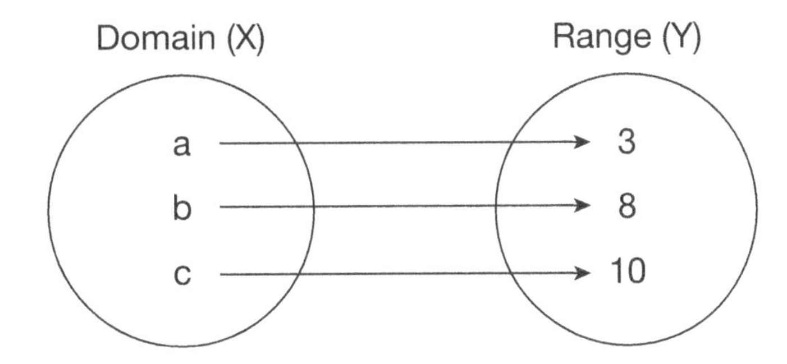

Not a Function: If the domain (x) is repeating, it's not a function. You do not need to check the range (y). It does not matter if the range repeats.

Not a Function

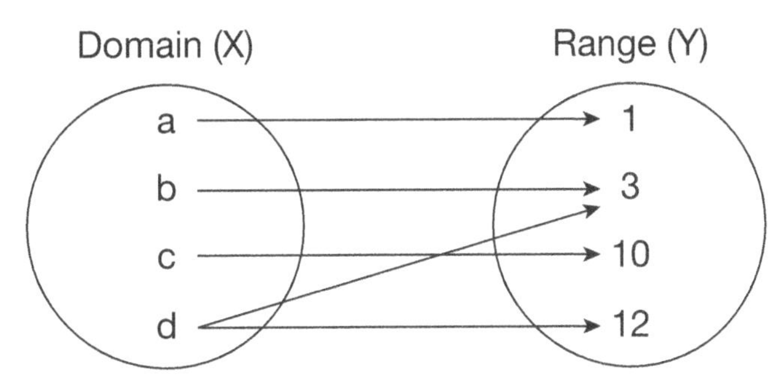

This is not a function because the domain (x) repeats.

Examples:

Function

Domain (x)	Range (y)
1	3
2	5
3	12
4	15
5	20

Not a Function

Domain (x)	Range (y)
1	0
2	2
1	4
3	3
5	2

FUNCTIONS

Examples:

(a, 1), (b, 3), (c, 5) is a function because the domains do not repeat.

(a, 1), (b, 3), (c, 5), (a, 10) is not a function because the domains repeat.

Types of graphs which are functions

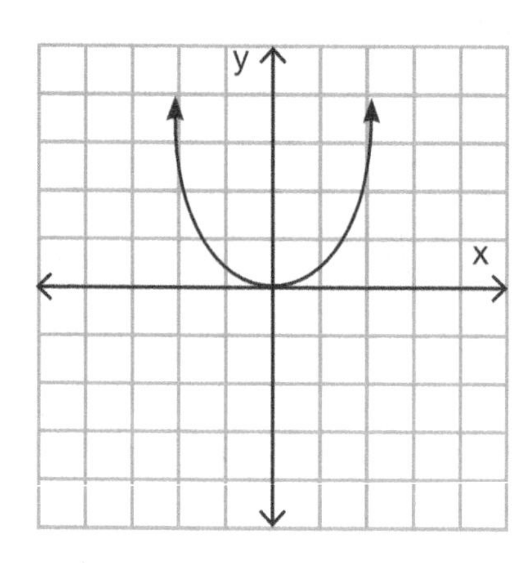

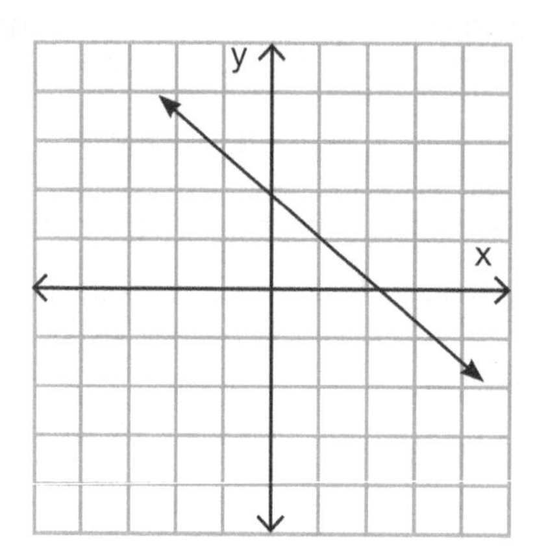

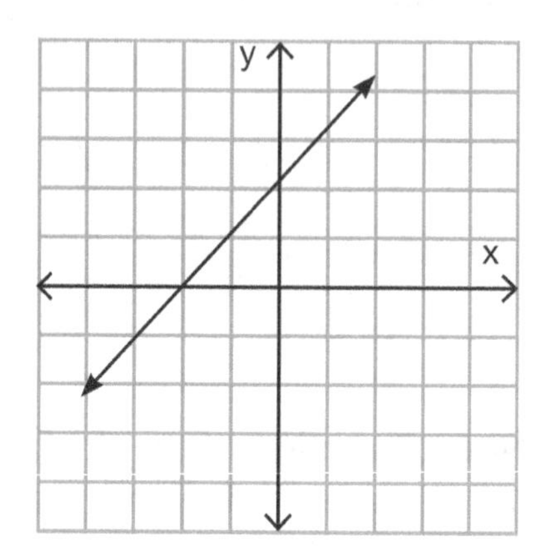

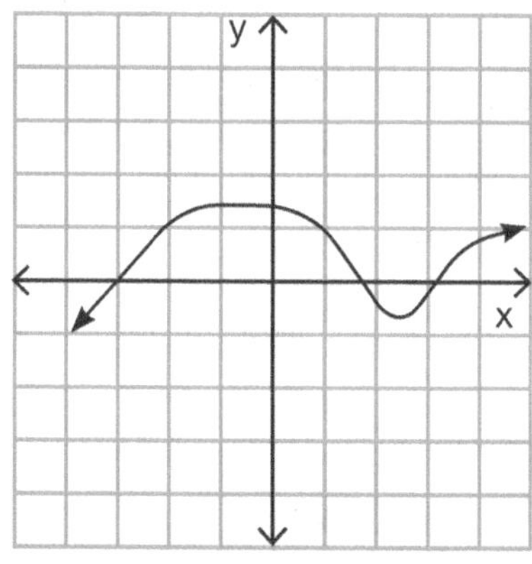

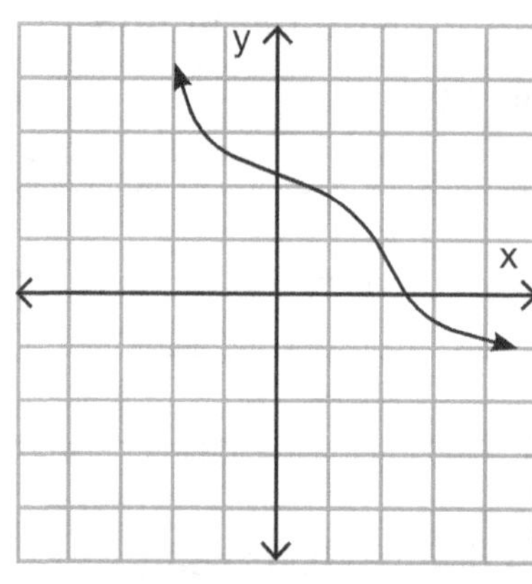

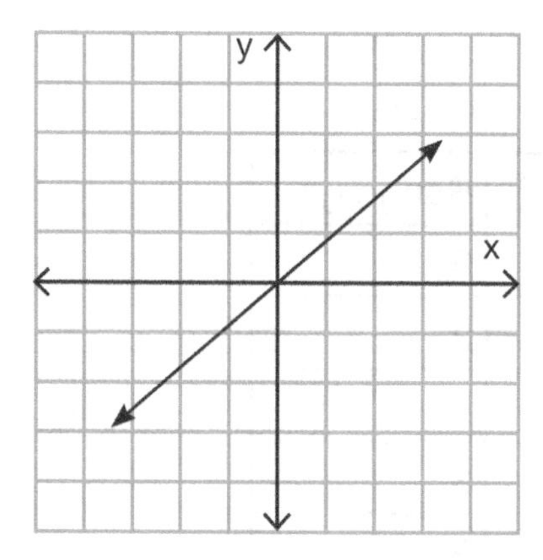

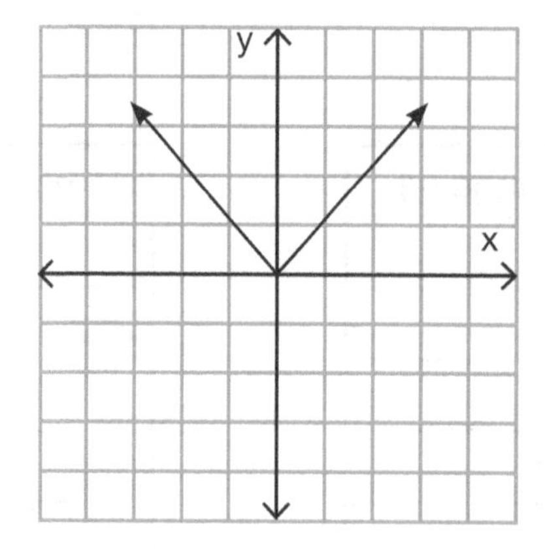

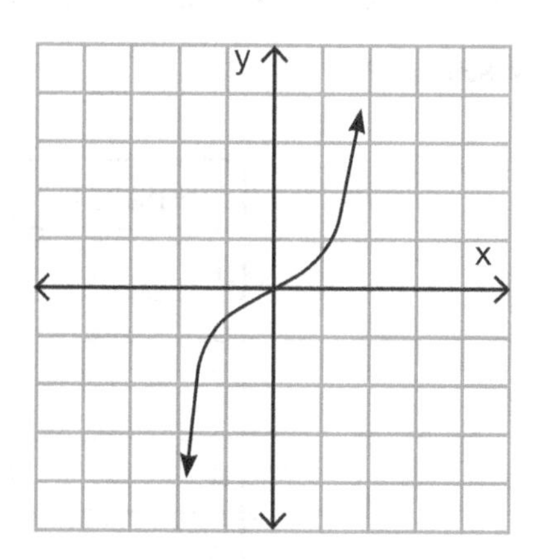

FUNCTIONS

Types of graphs which cannot be functions

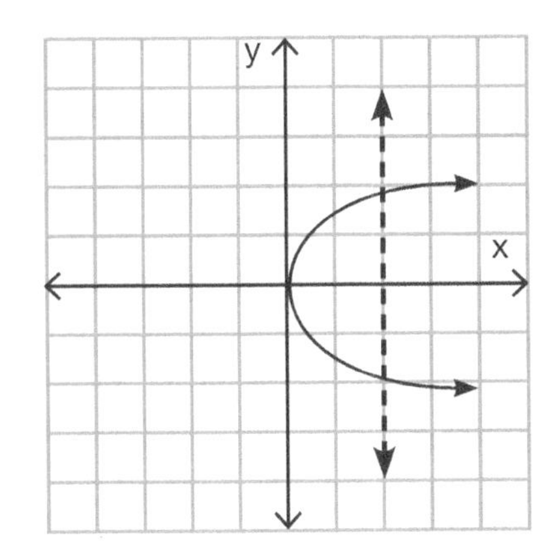

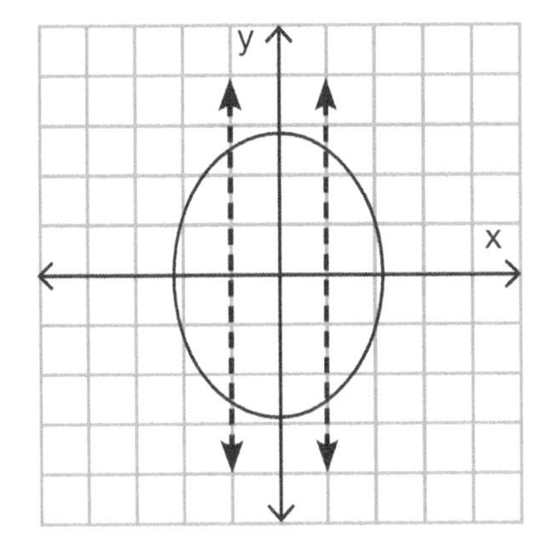

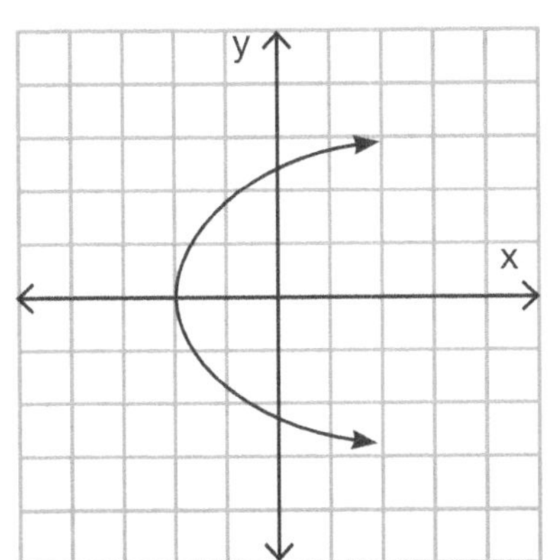

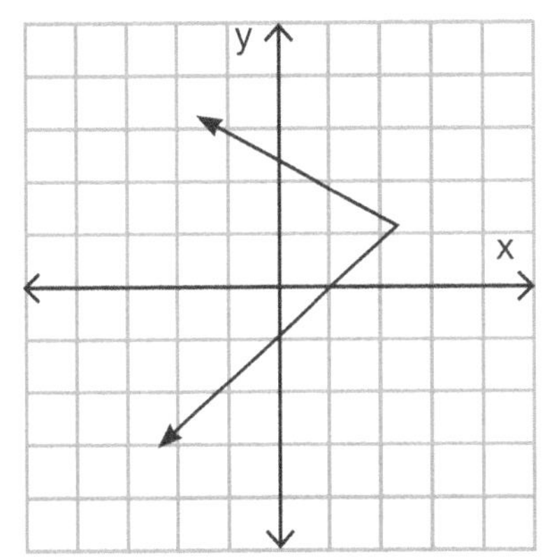

Function Operations

1. **Addition:** $(f+g)(x) = f(x) + g(x)$

   Example: $f(x) = 4x-2$ and $g(x) = 2x-1$, then $(4x-2)+(2x-1) = 6x-3$

2. **Subtraction:** $(f-g)(x) = f(x) - g(x)$

   Example: $f(x) = 4x-2$ and $g(x) = 2x-1$ then $(4x-2)-(2x-1) = 2x-1$

3. **Multiplication:** $(f \times g)(x) = f(x) \times g(x)$

   Example: $f(x) = 4x-2$ and $g(x) = 2x-1$ then $(4x-2) \times (2x-1) = 8x^2 - 8x + 2$

4. **Division:** $(f \div g)(x) = f(x) \div g(x)$

   Example: $f(x) = 4x-2$ and $g(x) = 2x-1$ then $(4x-2) \div (2x-1) = \dfrac{4x-2}{2x-1} = \dfrac{2(2x-1)}{2x-1} = 2$

Functions Worksheet

Determine if the following relations are functions. Then, determine the domain and range.

1. (1, 2), (9, 13), (0, 7), (5, 13)

Function:__________

Domain:__________

Range:__________

2. (3, 5), (7, 8), (9, 4), (7, 19)

Function:__________

Domain:__________

Range:__________

Determine if the following relations are functions. Then determine the domain and range.

3.

x	y
3	2
5	4
6	7
11	8

Function:__________

Domain:__________

Range:__________

4.

x	y
a	13
b	14
c	15
a	16

Function:__________

Domain:__________

Range:__________

Functions Challenge

1. Which of the following tables does not represent a function?

A)

x	F(x)
a	1
b	3
c	5

B)

x	F(x)
John	Smart
Peter	Intelligent
Peter	Genius

C)

x	F(x)
2	1
l	3
m	1

D)

x	F(x)
m	4
n	5
k	4

2. Which of the following tables does not represent a linear relation?

A)

X	Y
1	0
2	2
3	4
4	6

B)

X	Y
−1	−3
−2	−6
−3	−9
−4	−12

C)

X	Y
2	6
4	12
6	18
8	24

D)

X	Y
1	2
3	0
5	3
7	6

3. For the below function, find $f(2) - f(1) + 2f(3)$.

x	f(x)
1	7
2	8
3	9
4	10
5	11
6	12

A) 19 B) 26 C) −19 D) −26

Functions Challenge

4. If $g(x+1) = \frac{2x-3}{3x+4}$, then find g(2).

A) 7 B) $-\frac{1}{7}$ C) $\frac{1}{7}$ D) −3

5. If g(x) = 4x + 5, then what is the value of g(1) + g(−2)?

A) 1 B) 3 C) 6 D) 10

6. Which of following graph represents a function?

A)
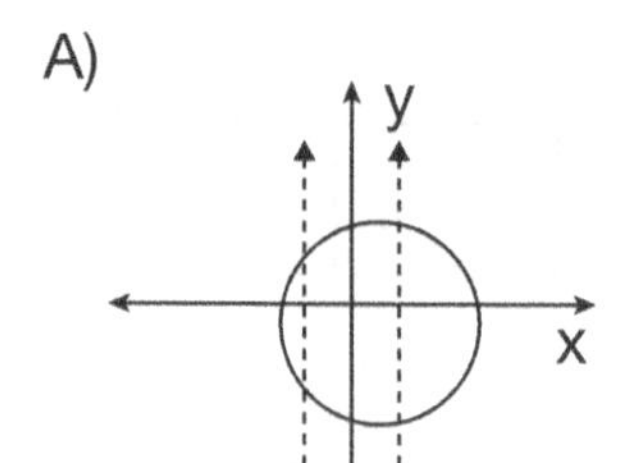

B)
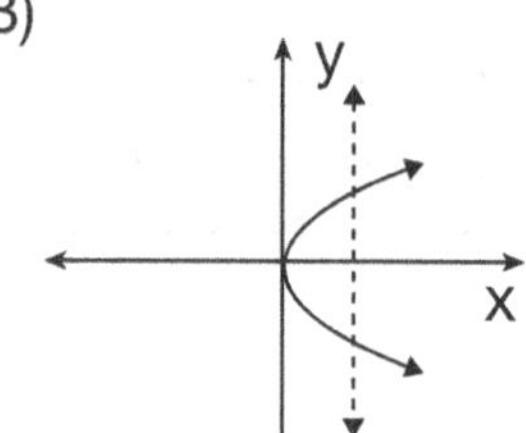

C)
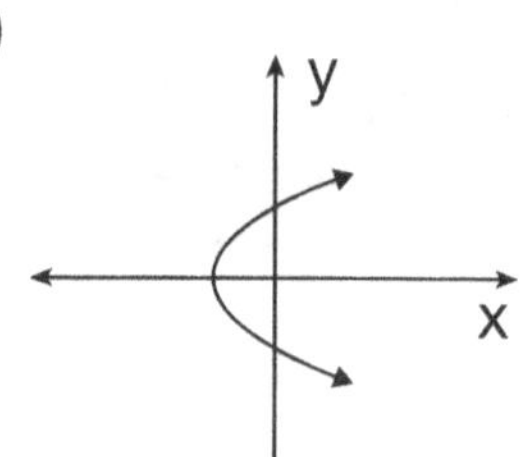

D)
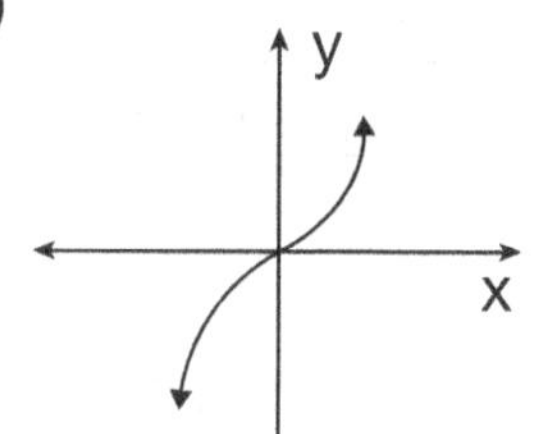

7. $f(x) = |x-4| - |2x|$, then find f(−2) + f(0) + f(3).

A) −1 B) 0 C) 1 D) 2

8. Determine which of the relations below are functions.

A) (1,2),(3,4),(4,6),(3,7) B) (a,2),(b,4),(c,6),(b,7) C) (m,2),(n,4),(k,6),(m,7) D) (1,2),(3,4),(7,8),(10,4)

LINEAR EQUATIONS & SLOPE

Definiton of a Linear Equation

Linear equation with two variables, can be written in the form $Ax + By + C = 0$.

Example:

$5x + 6y = 10$

Standard Form

Ax + By = C

When A, B and C are real numbers and A and B are not equal to zero, A must be positive.

Example:

$2x + 7y = 15$

Slope

Suppose there are two points on a line, (x_1, y_1) and (x_2, y_2).

The slope m of the line is:

$$m = \frac{\text{change in } y \text{ (Rise)}}{\text{change in } x \text{ (Run)}} = \frac{y_2 - y_1}{x_2 - x_1}$$

Example:

Find the equation of the line passing through the point (1, 2) and (3, 6)

Solution:

$$\text{Slope} = m = \frac{6-2}{3-1} = \frac{4}{2} = 2$$

Slope-Intercept Form

$y = mx + b$

m = slope

b = y–intercept = $(0, b)$

Point - Slope Form

$y - y_1 = m(x - x_1)$

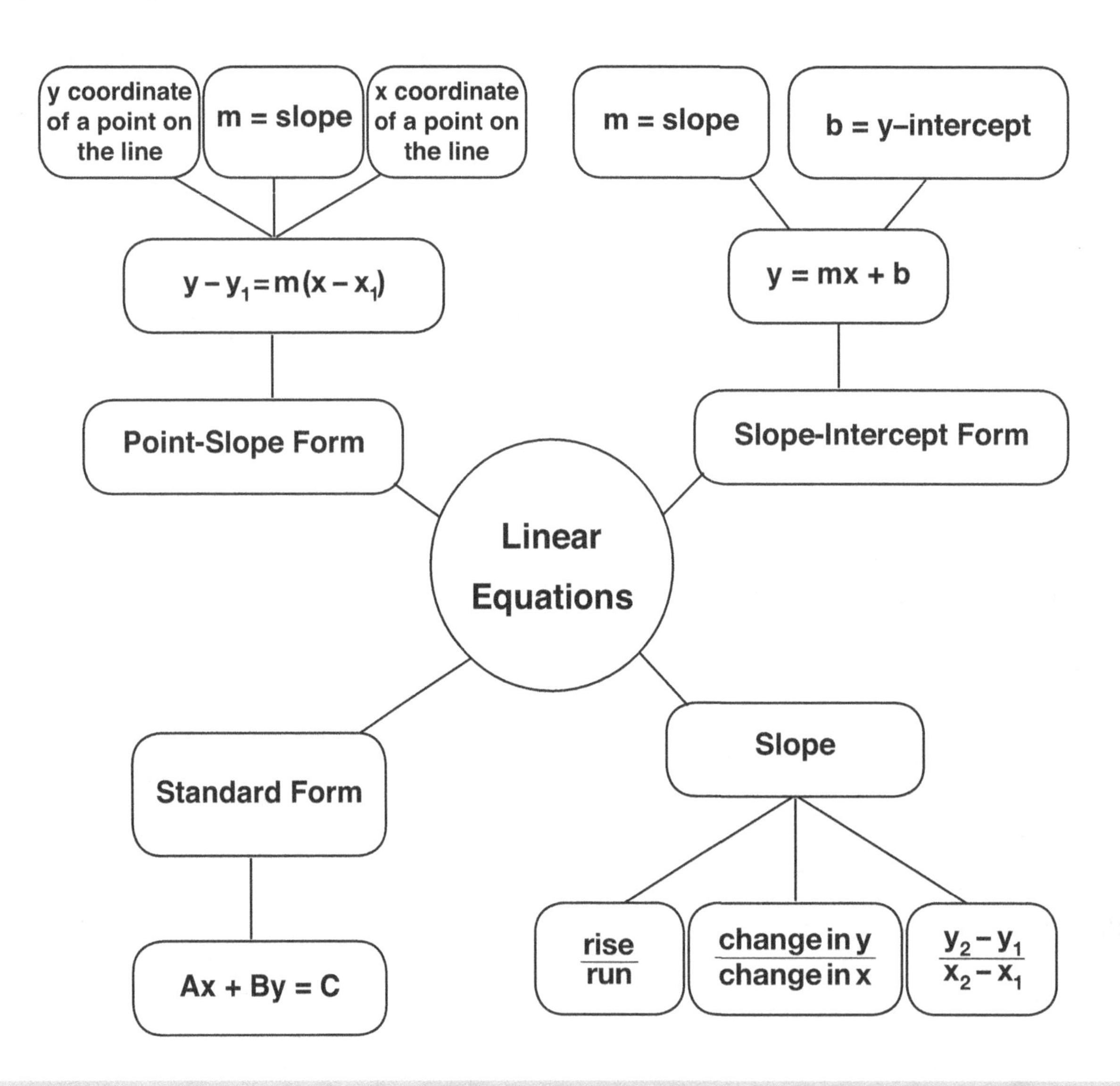

Linear Equations & Slope Worksheet

Find the slope of the line that passes through each pair of points. $\left[\text{Slope} = \frac{y_2 - y_1}{x_2 - x_1}\right]$

1. (3, 4) and (7, 9)

2. (2, 1) and (6, 8)

3. (4, 9) and (3, 7)

In each linear equation, find the slope and the y–intercept.

4. $y = 3x - 24$

5. $y = \frac{1}{3}x - 17$

6. $y = \frac{3}{4}x - \frac{1}{4}$

Find the equation of the line in slope–intercept form. ($y = mx + b$)

7. $m = 3$ and y intercept $= -19$

8. $m = 6$ and $(-1, 3)$

9. Write the equation of the line in slope–intercept form given the graph below.

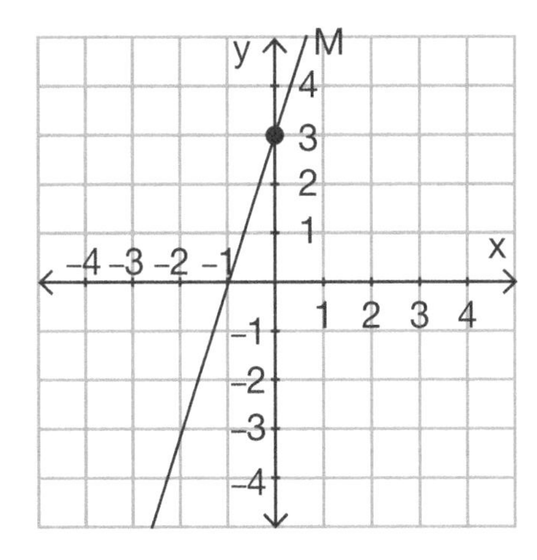

Slope:____________

y–Intercept:________

Equation:_________

Linear Equations & Slope Challenge

1. What is the value of k if a line passing through (6, 3) and (2,k) has a slope of −1?

A) 3 B) 5 C) 7 D) 9

2. What is the slope of the line that passes through (3, 2) and (1, 8)?

A) 2 B) −2 C) 3 D) −3

3. Which of the following equations is a line whose slope is 4 and passes through the point (3, 5)?

A) $y - 5 = 4(x - 3)$ B) $y - 5 = 4(x - 4)$ C) $y - 5 = 5(x - 3)$ D) $y - 3 = 4(x - 5)$

4. Which of the following equations is a line that passes through the coordinates (0, 3) and (3, 2)?

A) $y = -2x$ B) $y = 3x - 1$ C) $y = -\frac{1}{3}x + 3$ D) $y = \frac{1}{3}x + 3$

5. Which of the following is the slope of the graph?

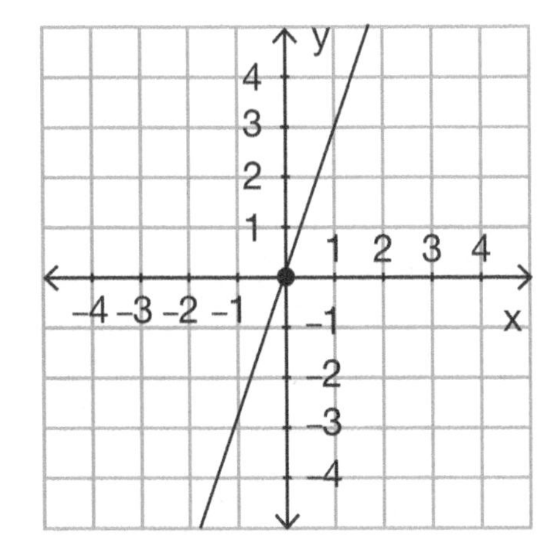

A) 3 B) −3 C) $\frac{1}{3}$ D) 0

6. From the graph below, which two points have the same slope in reference to the origin?

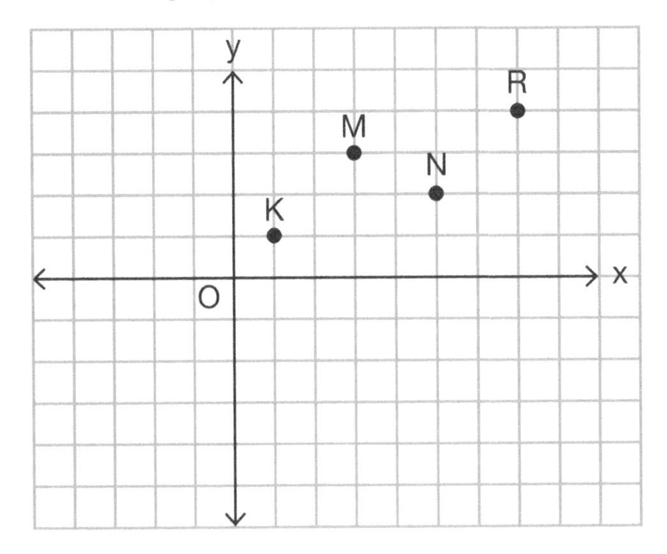

A) R and M B) M and K C) R and N D) N and K

7. What is the solution to the equation below?

$4\left(\frac{x-1}{8}\right)=\frac{4}{6}+3x$

A) $\frac{6}{7}$ B) $\frac{7}{6}$ C) 1 D) $-\frac{7}{15}$

8. Line M passes through the coordinate points $\left(-2,\frac{1}{3}\right)$ and $\left(1,\frac{-2}{6}\right)$. What is the slope of line M?

A) $\frac{1}{9}$ B) $-\frac{2}{9}$ C) $-\frac{9}{2}$ D) -9

9. Which of the following equations represents a line that is perpendicular to the line that passes through points A(3, 6) and B(4, 8)?

A) $y = 2x - 9$ B) $y = -\frac{1}{2}x + \frac{15}{2}$ C) $y = 2x + 3$ D) $y = x + 8$

10. If the line L passes through the coordinate points (−1, 5) and (2, 8), what is the slope of line L?

A) 1 B) −1 C) 2 D) −2

UNIT RATE & PERCENTAGES

Distance = rate$\times$time

$D = r\times t$

Distance = mile / miles

Rate: miles per hour (mph)

Time: hour / hours

Example: If Vera's car goes 27 miles per gallon of gas, how far can go it on 9 gallons of gas?

Solution:

27 miles take 1 gallon

x miles takes 9 gallon

$x = 27\times 9 = 243$ miles.

Percent: A fraction that is a portion of 100.

$$\text{Perecent change} = \frac{\text{amount of change}}{\text{original amount}}$$

Example: $20\% = \frac{20}{100}$

3 types of percent questions

Examples:

1. What is 20% of 40?

Solution: $P=\frac{20\times40}{100}=8$

2. What percent of 30 is 120?

Solution: $\frac{n\times30}{100}=120$ then, $n=400$

3. 15 is 25% of what number?

Solution: $15=\frac{25\times m}{100}$ then, $m=60$

Example: If Vera's car goes 30 miles per gallon of gas, how far can it go on 12 gallons of gas?

Solution:

30 miles takes 1 gallon

x miles takes 12 gallons

$x = 30\times12 = 360$

Unit Rate & Percentages Worksheet

Use the data in each table below to find the unit rate.

1.

Days	1	2
Miles	30	60

Unit Rate: ________________ miles/day

2.

Hours	5	10
$	20	40

Unit Rate: ________________ $/hours

3.

Book	7	21
Pages	35	105

Unit Rate: ________________ pages/book

Solve each problem for the unknown.

4. 10 is 25% of what number?

5. What is 45% of 40?

6. What % of 30 is 6?

7. What is 10% of 90?

Unit Rate & Percentages Challenge

1. Melisa can solve 108 questions in 9 day.
How many questions can Melisa do in one days.

A) 9 B) 12 C) 18 D) 36

2. If Tony's car can drive 480 miles on a full tank of 16 gallons, what is the miles per gallon of his car?

A) 20 mpg B) 12 mpg C) 25 mpg D) 30 mpg

3. If Vera's car can drive 180 miles on a full tank of 9 gallons, then find the miles per gallon.

A) 20 mpg B) 25 mpg C) 30 mpg D) 35 mpg

4. A cleaning company charges $480 to clean 24 classrooms.
What is the company's price for cleaning a single class?

A) $10 B) $15 C) $20 D) $25

5. Vera bought 5 CDs for $60. What was the unit price?

A) $10 B) $12 C) $15 D) $20

6. Vera ran the 1km in 10 minutes. How fast did she run in meters per second?

A) 3 B) 5 C) $\frac{3}{5}$ D) $\frac{5}{3}$

7. A microwave originally priced at $40 is decreased in price by 25%. What is the sale price?

A) $40 B) $35 C) $30 D) $25

8. The price of a book has been discounted 20%. The sale price is $45. What is the original price?

A) $45.5 B) $51.5 C) $51.75 D) $56.25

9. What percent of 25 is 20?

A) 40% B) 50% C) 60% D) 80%

10. $1\frac{1}{8}$

Write the above fraction as a percent.

A) 12% B) 125% C) 112.5% D) 175%

11. 30% of 70 equals what percent of 84?

A) 15% B) 25% C) 35% D) 40%

12. There are 8 students on Mr. Tong's math team. If 75% of the team members are girls, how many students on the math team are boys?

A) 1 B) 2 C) 3 D) 4

ANGLES

Point: A dot or location.

Line: A line extended forever in both directions

Line Segment: A part of a line with two endpoints

Ray: A line that starts at one point and continues on forever in one direction

Parallel Lines: Parallel lines never intersect and stay the same distance apart

Perpendicular Lines: Perpendicular lines are lines that intersect at a right angle

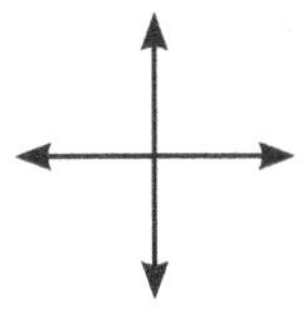

ANGLES

Intersecting Lines: Intersecting lines are two lines that share exactly one point

Acute Angle: An angle that is less than 90° but greater than 0°

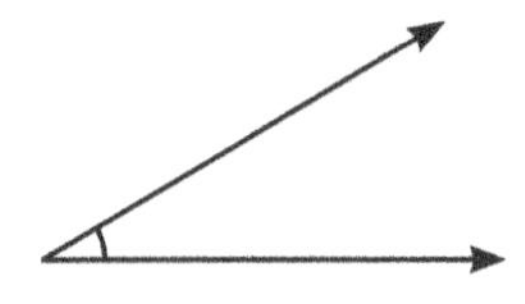

Right Angle: An angle that is exactly 90°

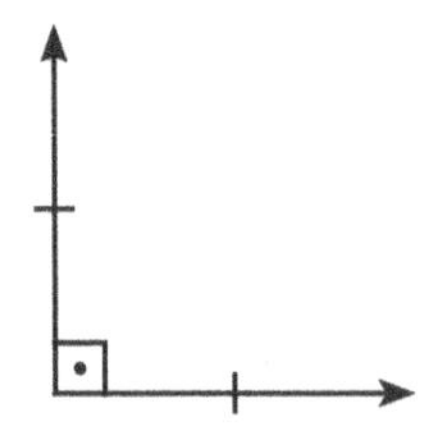

Obtuse Angle: An angle that is greater than 90° but less than 180°

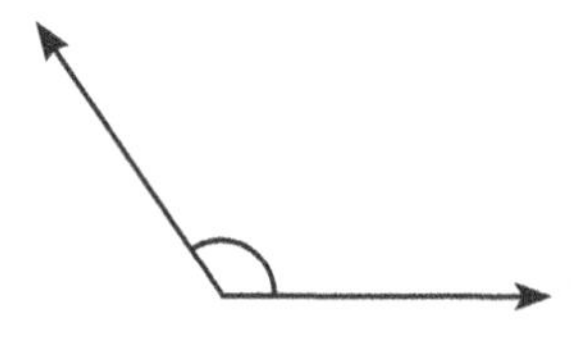

Straight Angle: An angle that is exactly 180°

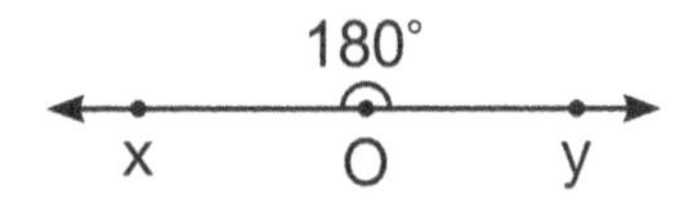

ANGLES

Corresponding Angles

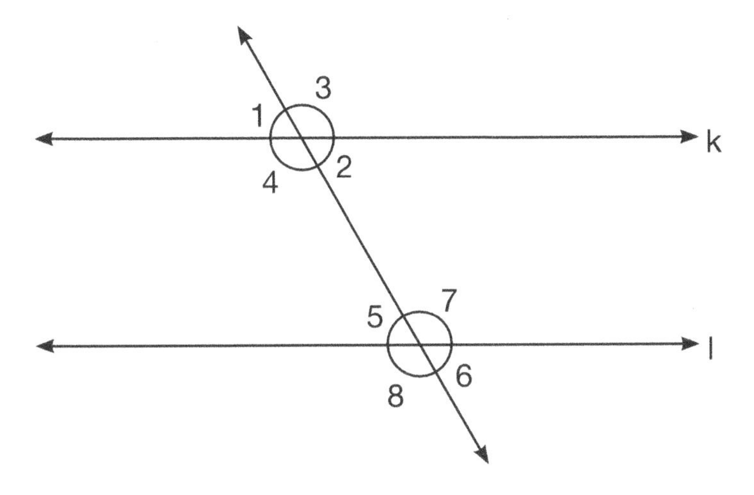

k // l

In the figure:

1 & 5, 3 & 7, 4 & 8, 2 & 6 are corresponding angles.

Alternate Interior Angles

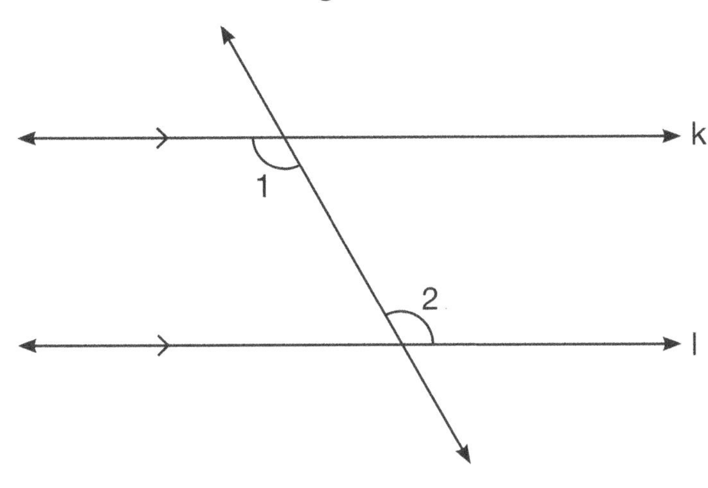

if k // l, then

$\angle 1 \cong \angle 2$

Alternate Exterior Angles

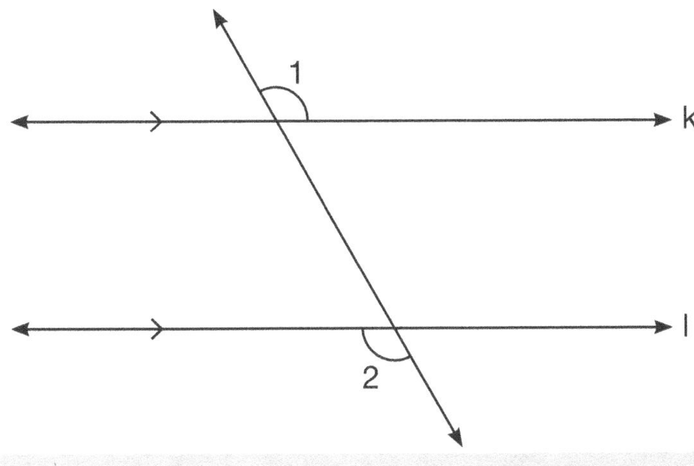

if k // l, then

$\angle 1 \cong \angle 2$

Angles Worksheet

Find the missing angle measures in each figure.

1.

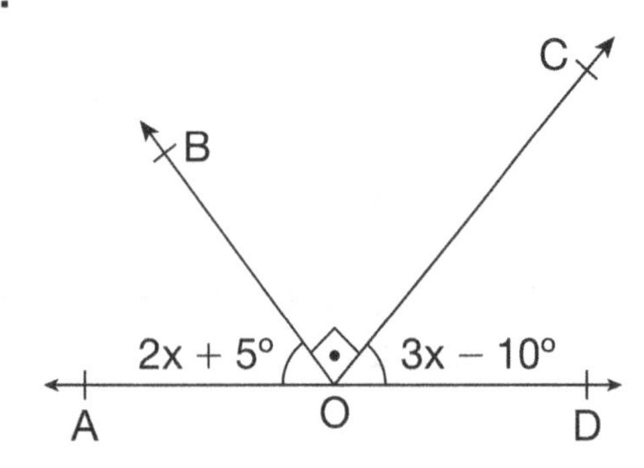

x = ______________

2.

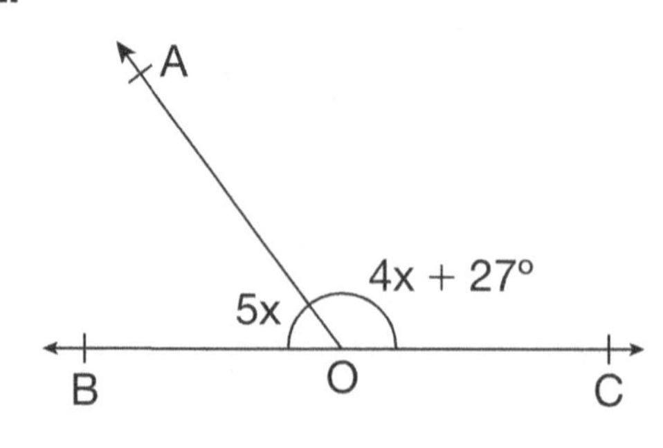

x = ______________

3.

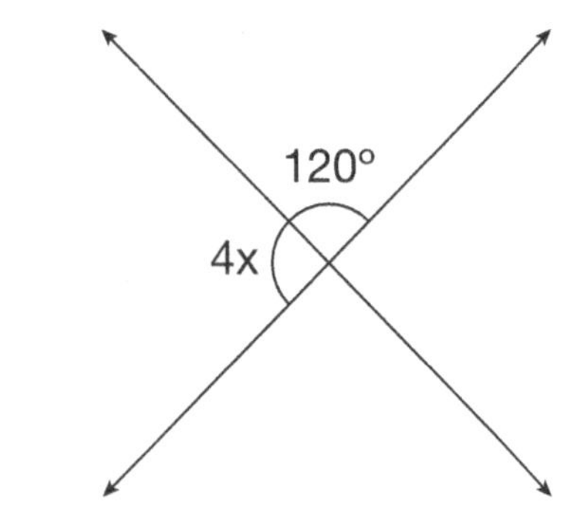

x = ______________

4.

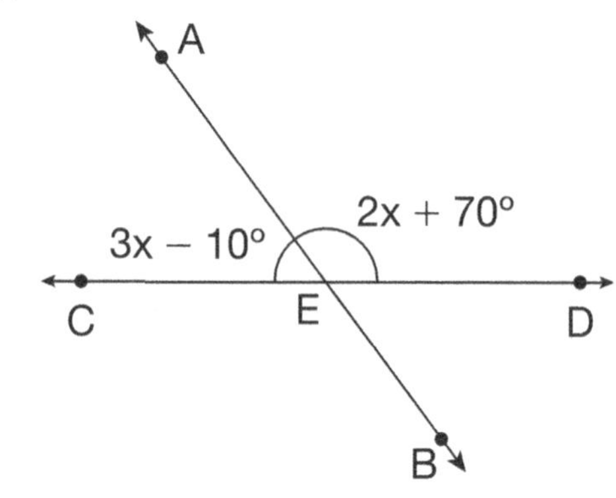

x = ______________

5. Given: AB // EG

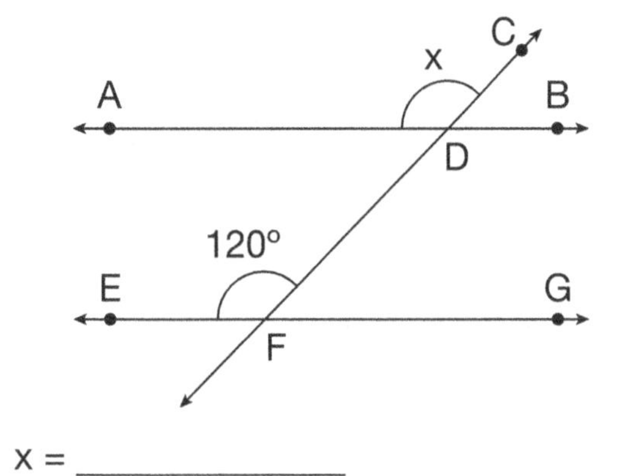

x = ______________

6.

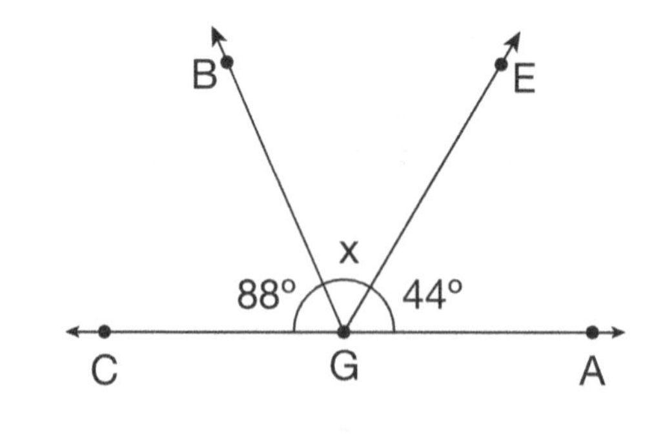

x = ______________

Angles Challenge

1. Find $\angle ABD$.

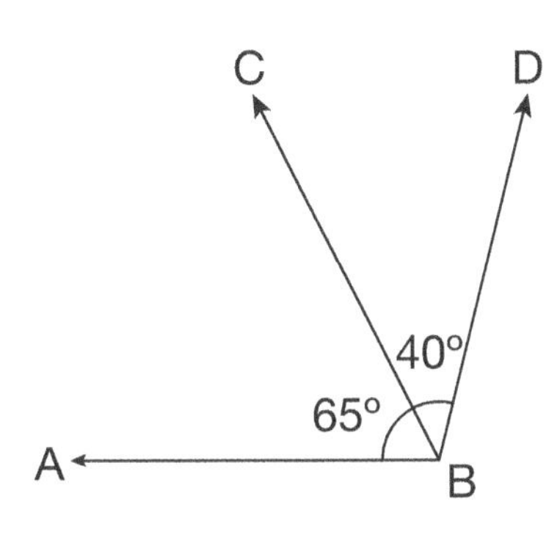

A) 105° B) 100° C) 95° D) 90°

2. Find x.

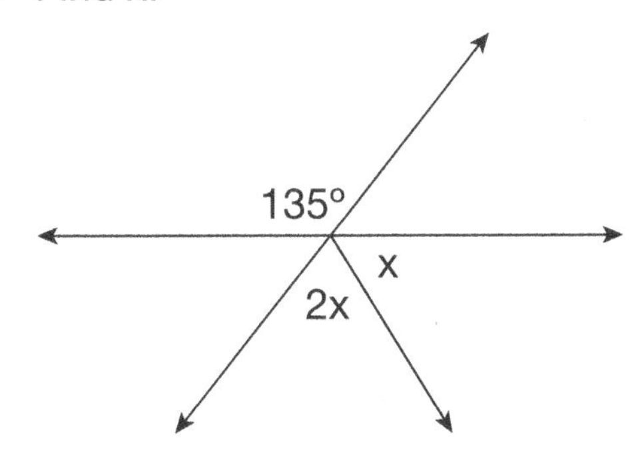

A) 10° B) 15° C) 25° D) 45°

3. Find x.

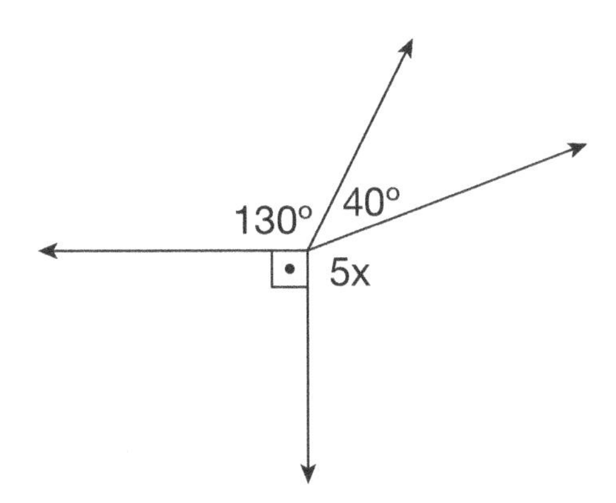

A) 30° B) 20° C) 10° D) 6°

4. Find x.

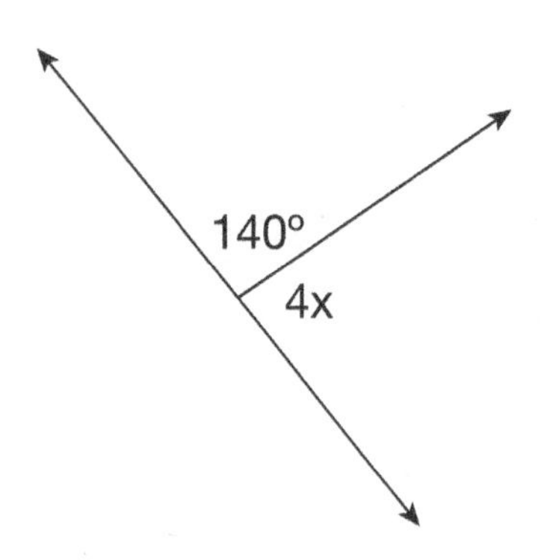

A) 40° B) 35° C) 20° D) 10°

5.

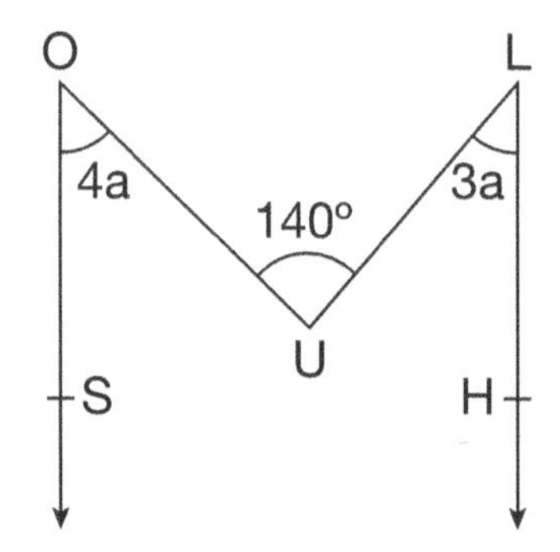

OS // LH

$m\angle SOU = 4a$, $\angle ULH = 3a$ and $\angle OUL = 140°$

So what is the value of 2a?

A) 30° B) 40° C) 50° D) 80°

6. What is the value of x?

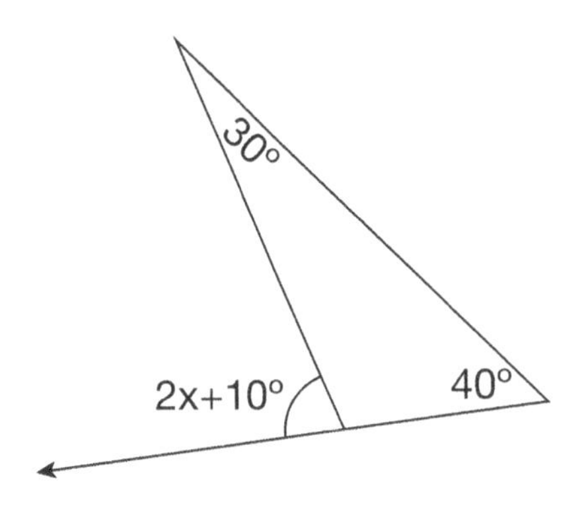

A) 10° B) 20° C) 30° D) 40°

7. What is the value of x?

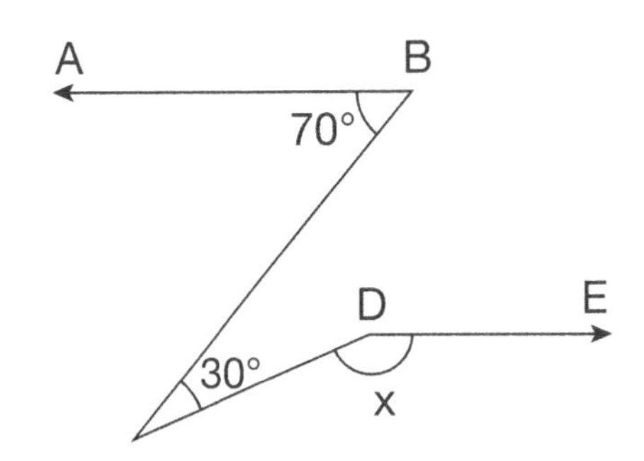

BA // DE

A) 120° B) 140° C) 150° D) 160°

8. From the figure ∠ BOC = 4x and ∠ AOD = 6x, What is the value of x?

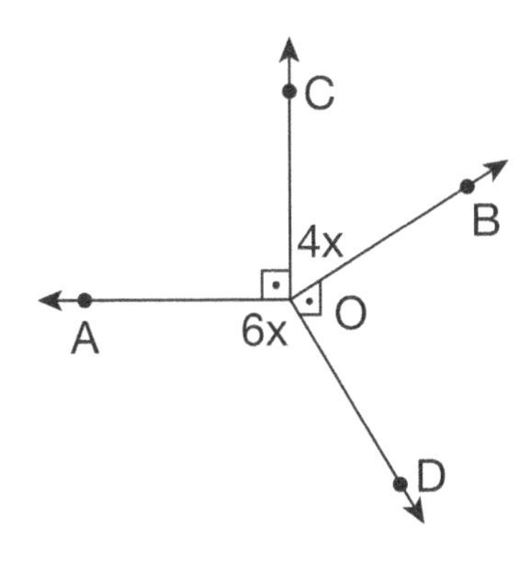

A) 10° B) 18° C) 30° D) 40°

9. In the figure below, AB is parallel to CD. What is the measure of $\angle$ BAE?

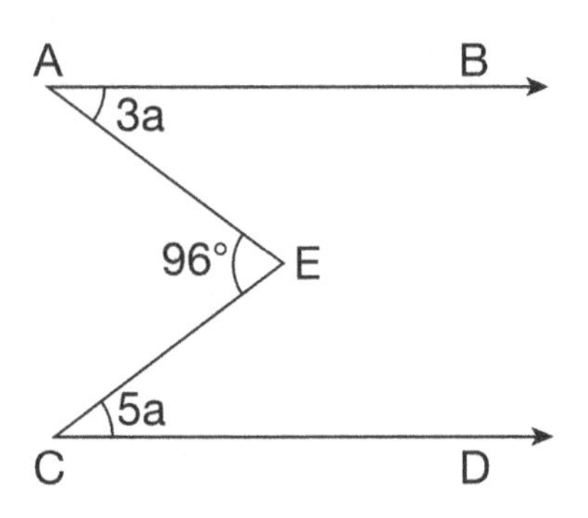

A) 10° B) 20° C) 25° D) 36°

10. What type of angle is represented in the diagram?

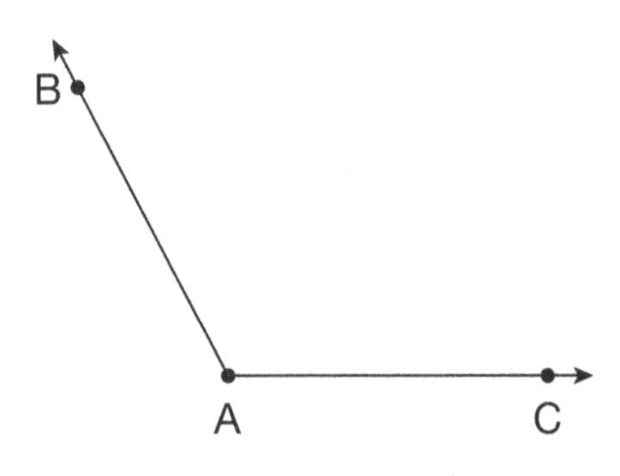

A) Obtuse B) Right C) Acute D) None

DISTANCE & MIDPOINT

The Distance Formula:

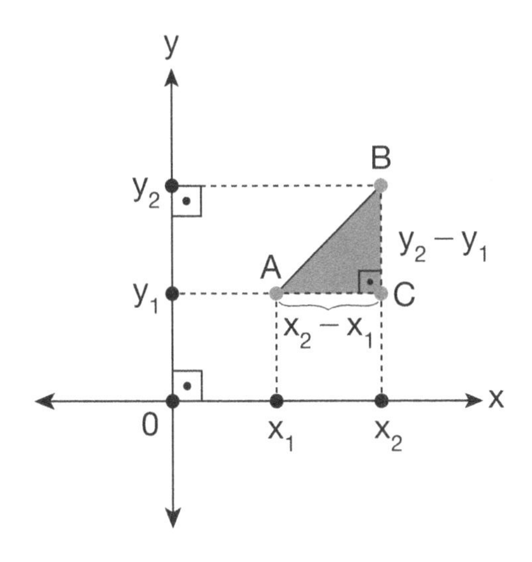

Use the Pythagorean theorem to find the length of $\overline{AB}$, which is the hypotenuse of the right triangle.

$$|AB|^2 = (x_2 - x_1)^2 + (y_2 - y_1)^2$$

$$|AB| = \sqrt{(x_2 - x_1)^2 + (y_2 - y_1)^2}$$

Midpoint formula:

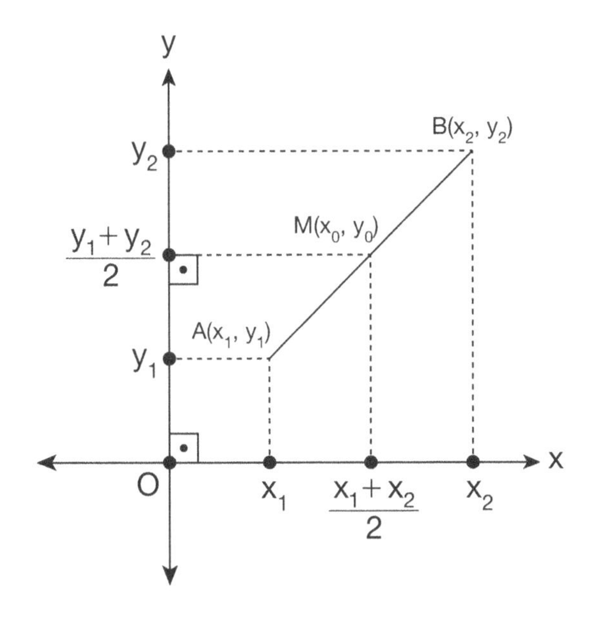

What are the coordinates (x_0, y_0) of the midpoint M of the segment whose endpoints are $A(x_1, y_1)$ and $B(x_2, y_2)$?

$$x_0 = \frac{x_1 + x_2}{2}$$

$$y_0 = \frac{y_1 + y_2}{2}$$

Distance & Midpoint Worksheet

Find the distance between each pair of points.

1. (3, 5) and (7, 8)

2. (−1, 3) and (−5, −8)

3. (6, 4) and (−3, 12)

4. $\left(\frac{1}{2}, 6\right)$ and $\left(\frac{1}{3}, 6\right)$

5. $\left(\frac{1}{2}, 2\right)$ and $\left(\frac{1}{2}, 8\right)$

6. $\left(\frac{1}{2}, 0\right)$ and $\left(-\frac{1}{4}, 0\right)$

Find the midpoint of the given points.

7. (0, 5) and (2, 9)

8. (−2, 4) and (−8, −8)

9. (7, 6) and (−1, 0)

10. $\left(\frac{1}{3}, 9\right)$ and $\left(\frac{1}{3}, 9\right)$

11. $\left(\frac{1}{6}, 1\right)$ and $\left(-\frac{1}{6}, 7\right)$

12. $\left(\frac{1}{8}, 0\right)$ and $\left(-\frac{1}{8}, 0\right)$

Distance & Midpoint Challenge

1. Find the distance between the following two points.

(9, 7) and (7, 3)

A) 5 B) $\sqrt{5}$ C) $2\sqrt{5}$ D) 10

2. Calculate the midpoint of the segment with the given endpoints:

(9, 7) and (7, 3)

A) (8, 3) B) (8, 5) C) (9, 7) D) (5, 8)

3. Find the perimeter of the triangle with vertices A(−8, −3), B(−2, −3), and C(−8, 5).

A) 10 units B) 14 units C) 22 units D) 24 units

4. On the coordinate plane below, what is the midpoint of AB?

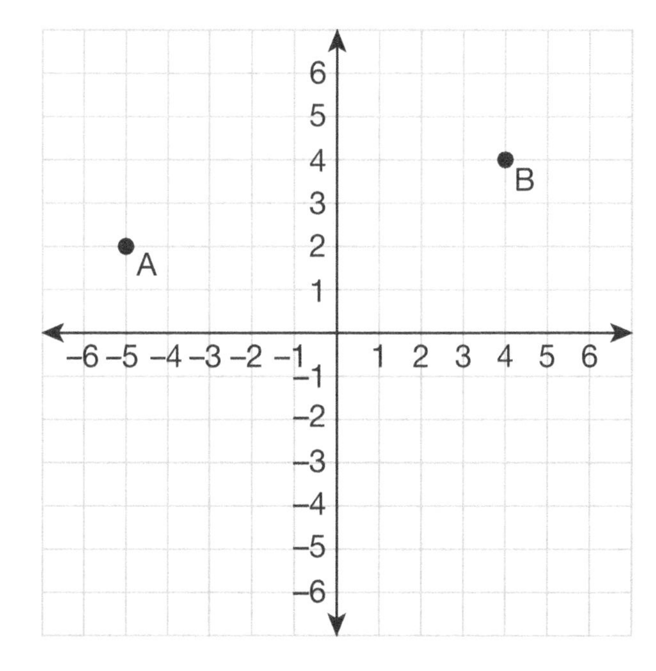

A) (0.5, −3) B) (0.5, 3) C) (−0.5, 3) D) (−0.5, −3)

Distance & Midpoint Challenge

5. The midpoint between the points A (4, 8) and B (a, b) is (12, 16). What are the coordinates of point B?

A) (20, 24) B) (24, 20) C) (–24, –20) D) (–20, –24)

6. AB has endpoints at 2 and 8. What are the coordinates of its midpoint?

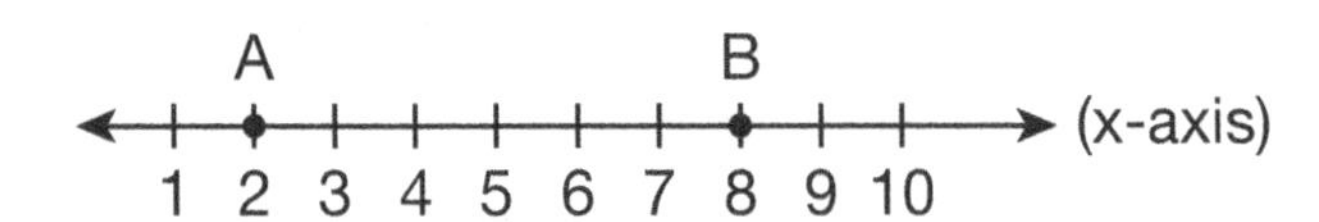

A) (3, 0) B) (5, 0) C) (7, 0) D) (9, 0)

7. Find the distance between (10, 14) and (2, 8).

A) 4 B) 5 C) 10 D) 15

8. Find the midpoint of (–1, –13) and (9, 23).

A) (4, 5) B) (4, –5) C) (–4, –5) D) (–4, 5)

Triangle Angles

The sum of the interior angles of a triangle is equal to 180°.

Example:

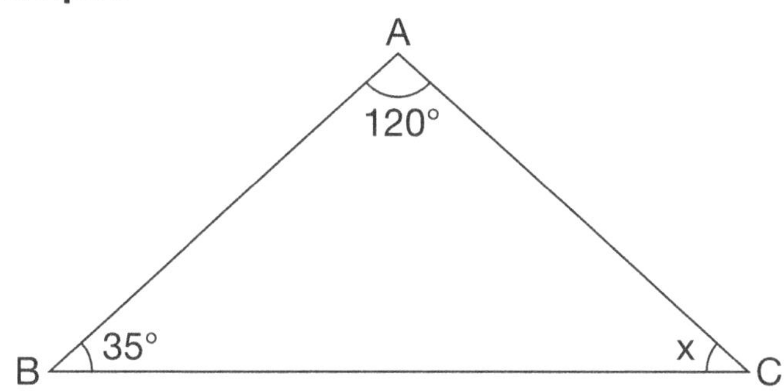

Solution: $35° + 120° + x = 180°$. Then, $x = 180° - 155° = 25°$

Types of Triangle

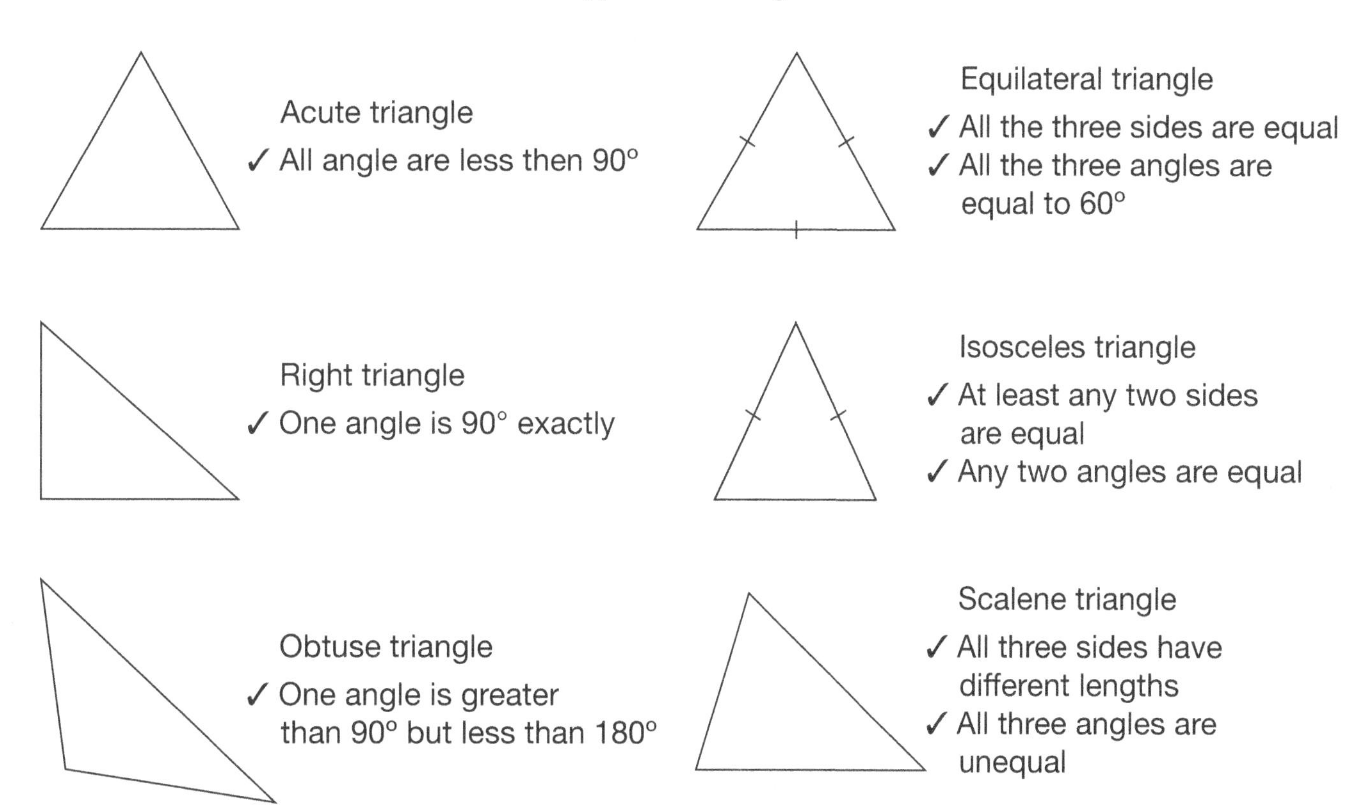

Special Right Triangles

▸ 30° : 60° : 90°

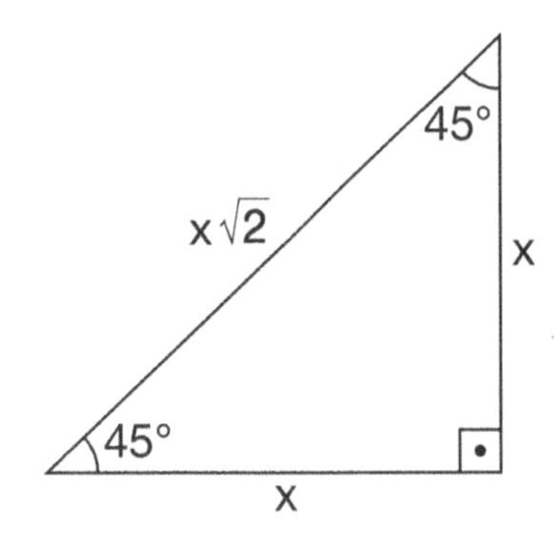

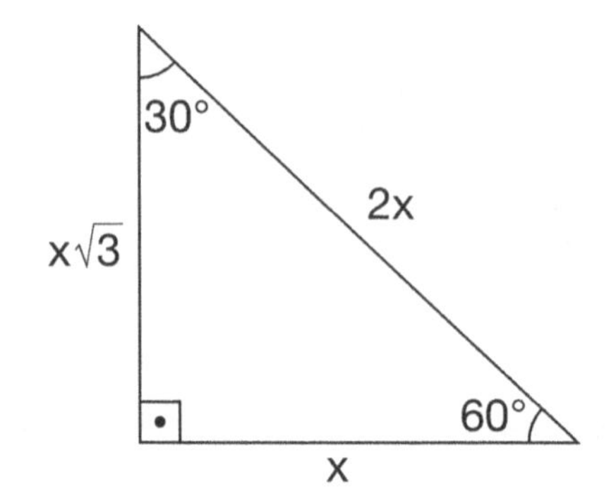

Equilateral Triangle

▸ 60° : 60° : 60°
(Equilateral Triangle)

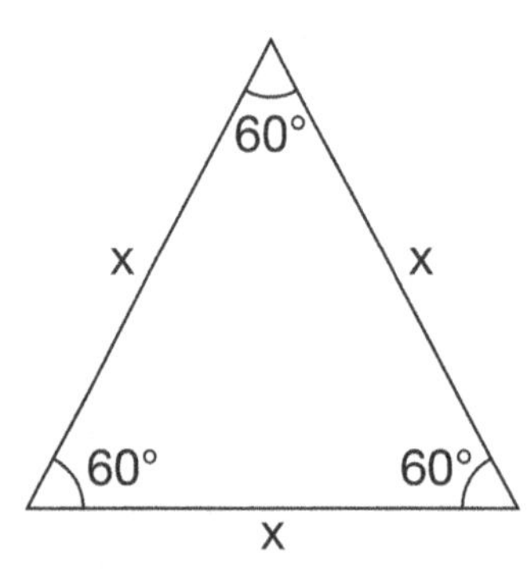

Triangles and Types of Triangles Worksheet

Identify each triangle based on both sides and angles.

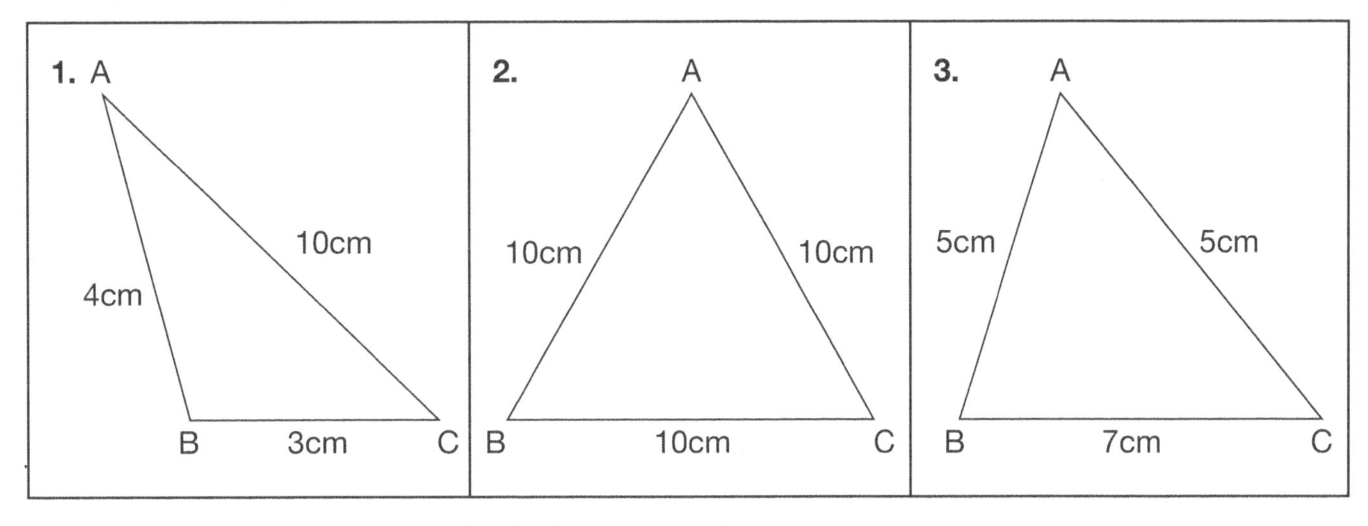

Calculate the degree of the angles in the triangles below

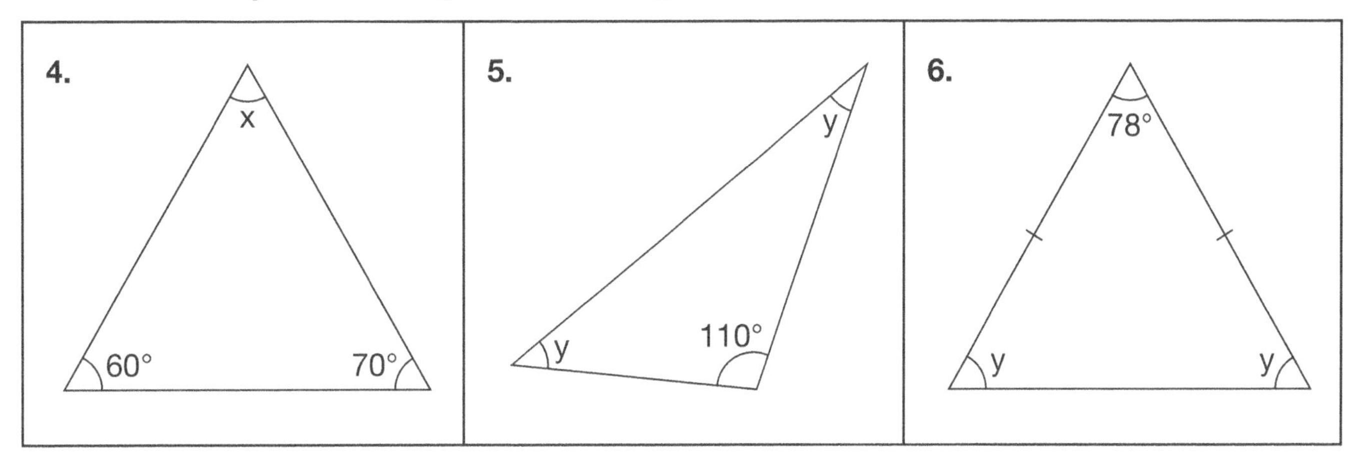

Find the each missing side lengths.

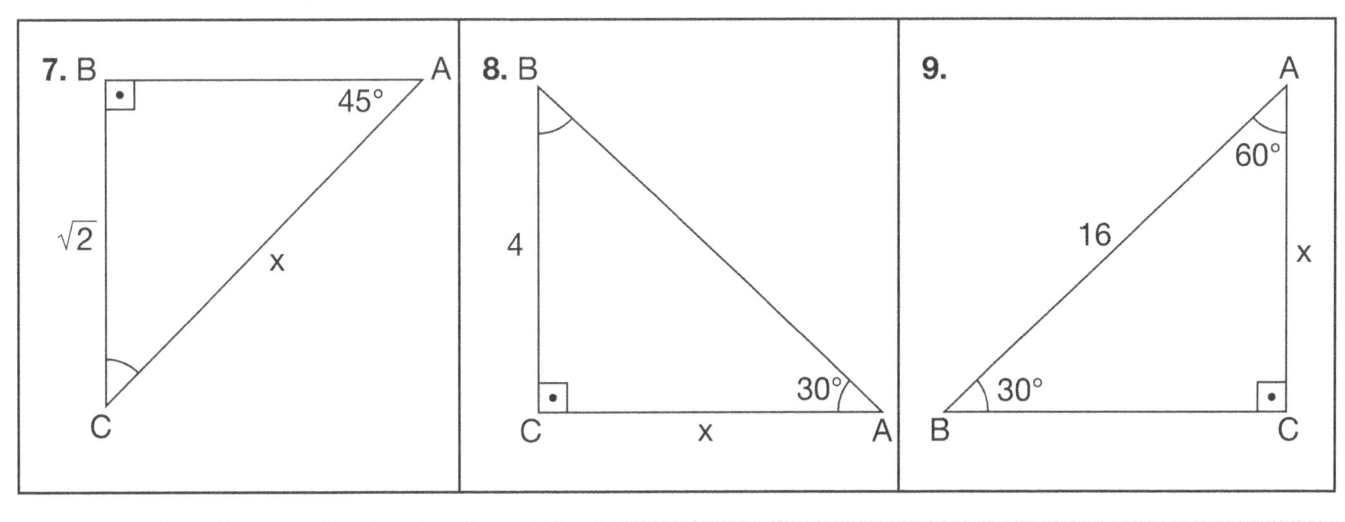

Triangles and Types of Triangles Challenge

1.

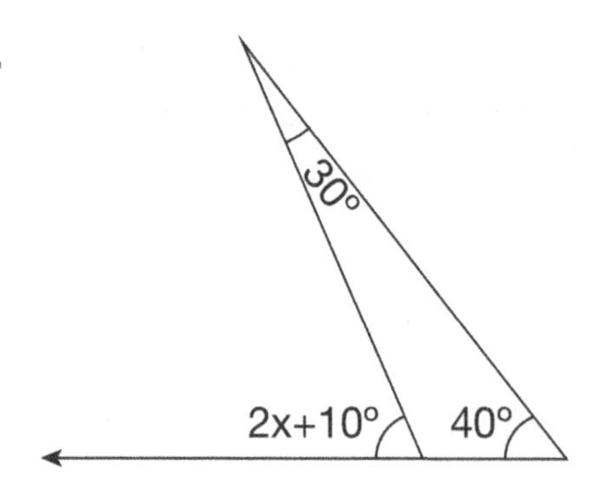

What is the value of x?

A) 10° B) 20° C) 30° D) 40°

2.

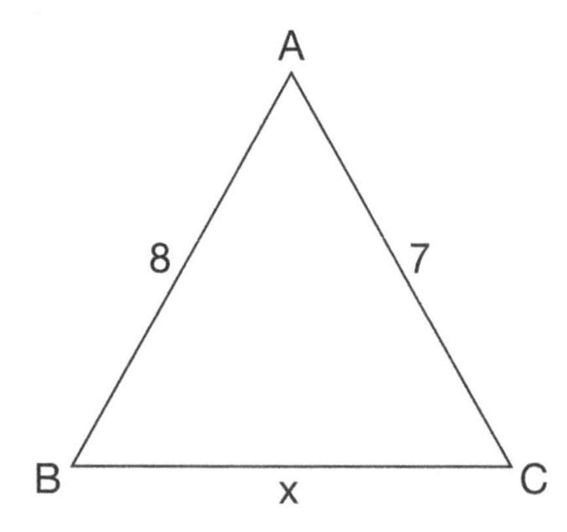

Which of following can be x?

A) 1 B) 15 C) 16 D) 12

3.

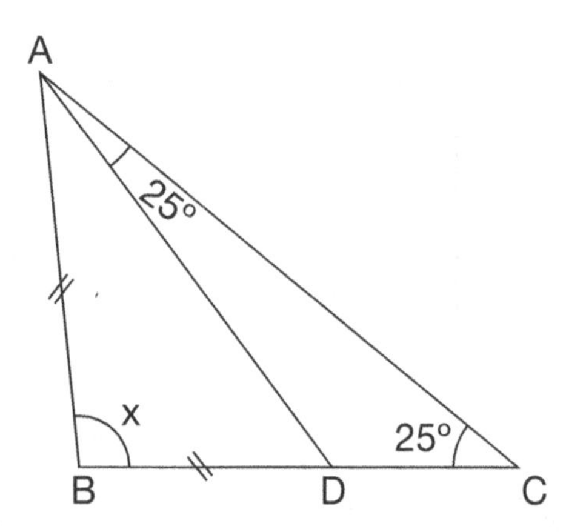

Find the value of x in the diagram

A) 60° B) 70° C) 80° D) 90°

4.

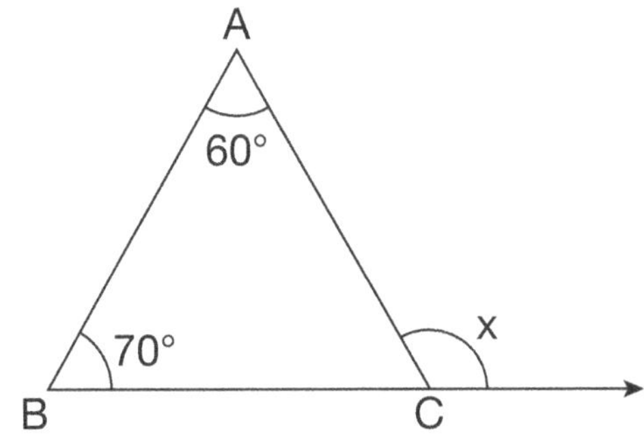

Find the value of x in the diagram

A) 110° B) 120° C) 130° D) 135°

5.

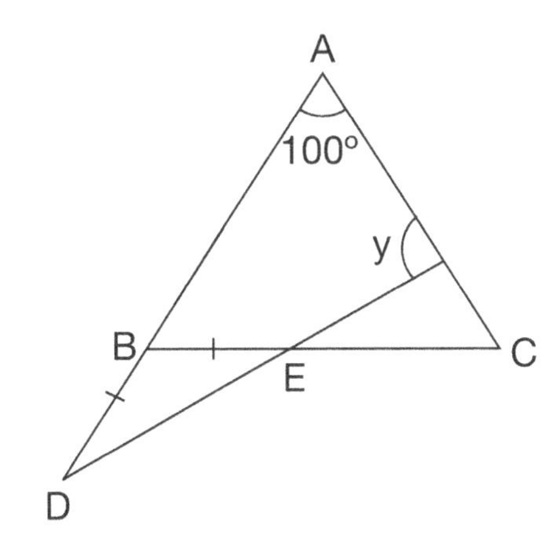

From the above figure if $|\overline{AB}|=|\overline{AC}|$ and $|\overline{BD}|=|\overline{BE}|$, what is the value of y?

A) 15 B) 30° C) 45° D) 60°

6.

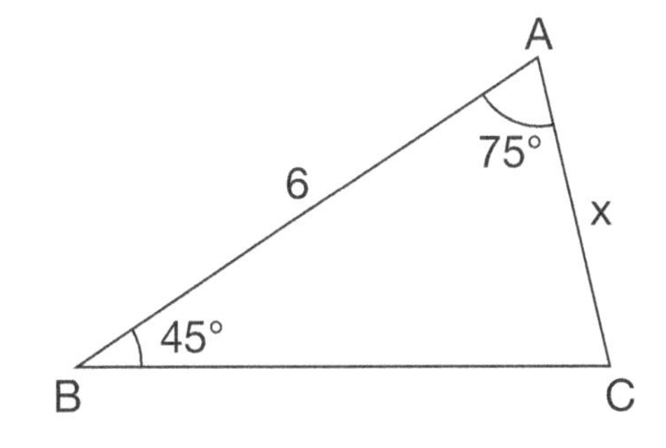

What is the value of x?

A) $\sqrt{6}$ B) $3\sqrt{2}$ C) $2\sqrt{6}$ D) 6

7.

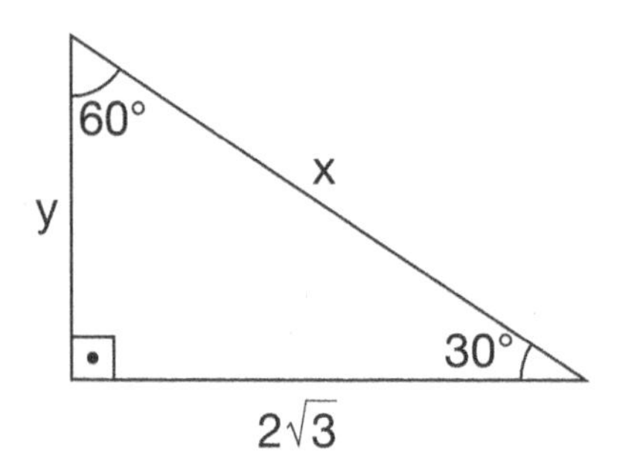

What are the values of x and y?

A) 2,4 B) 4,2 C) $\sqrt{3}$,2 D) 4,$\sqrt{3}$

8.

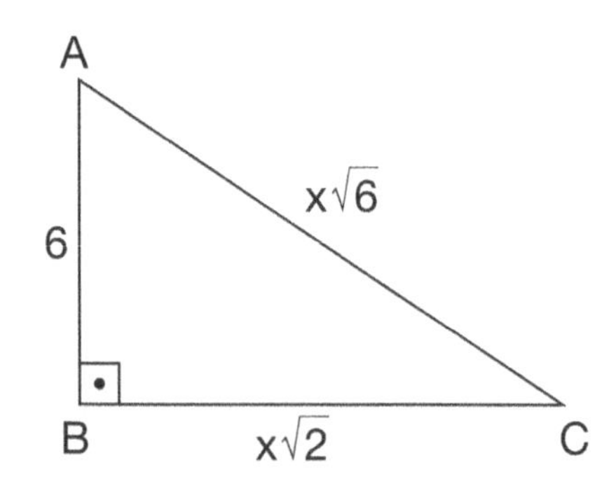

What is the value of x?

A) 1 B) 3 C) 5 D) 9

9.

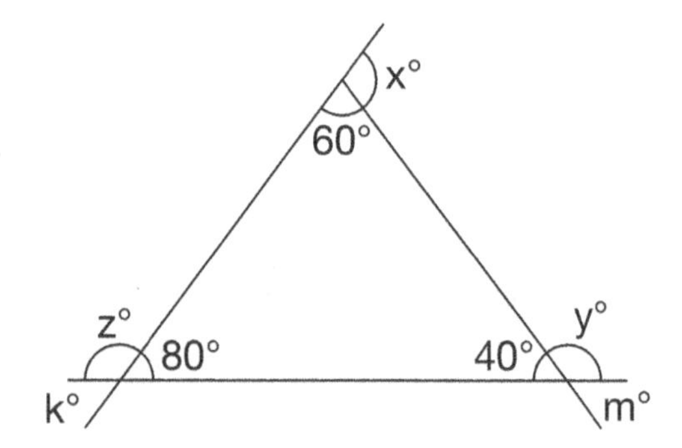

In the figure above, wich of the following is the greatest?

A) x° B) y° C) z° D) k°

10.

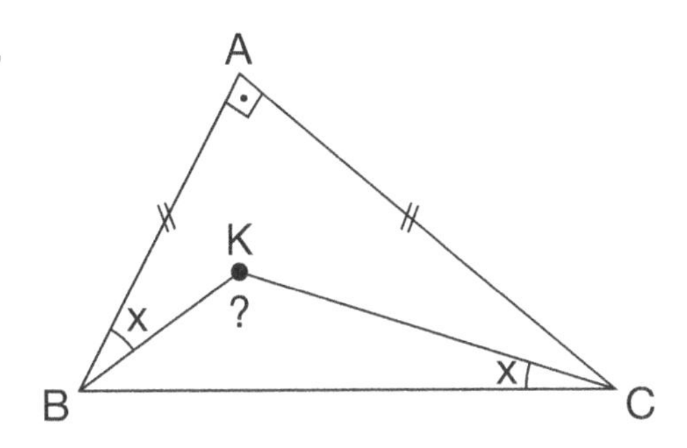

Find the measure of ∆ BKC.

A) 110°   B) 115°   C) 135°   D) 120°

11.

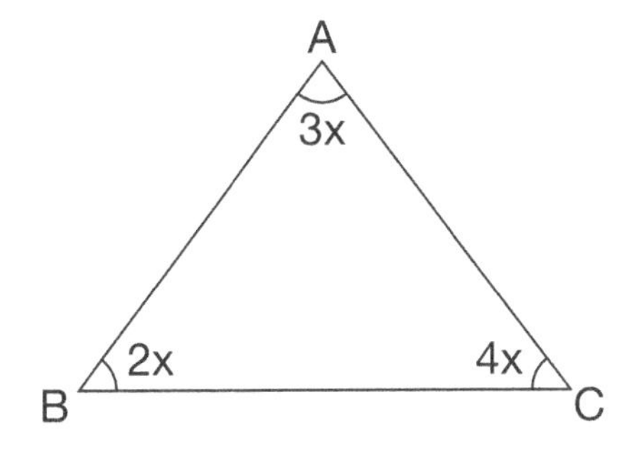

What is the value of x?

A) 10°   B) 20°   C) 30°   D) 40°

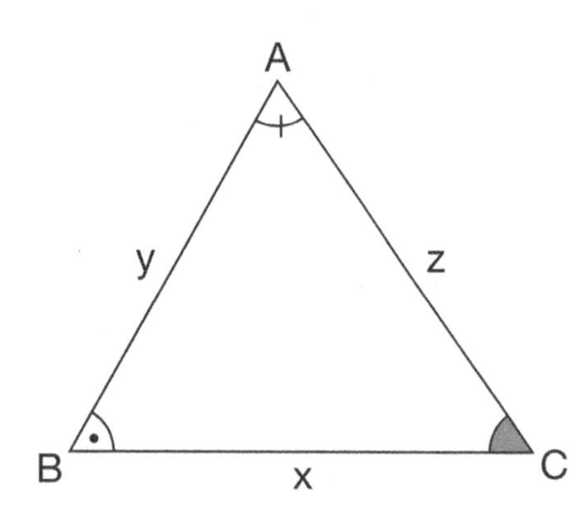

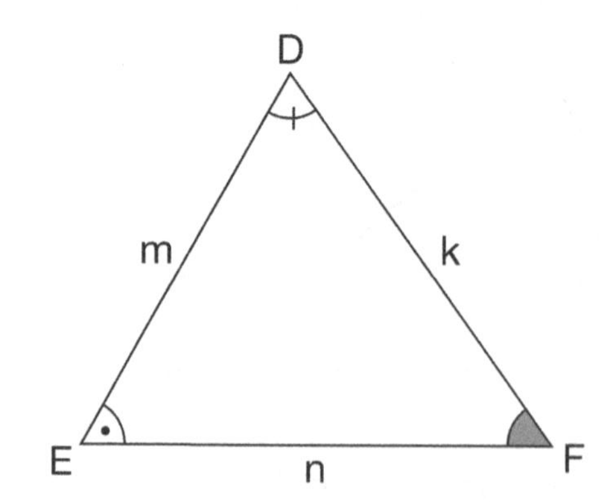

When two figures are similar, the ratios of the lengths of their corresponding sides are equal.

If $\widehat{ABC} \sim \widehat{DEF}$ then $\angle A = \angle D$ and $\dfrac{|AB|}{|DE|} = \dfrac{|AC|}{|DF|} = \dfrac{|BC|}{|EF|}$

$\angle B = \angle E$

$\angle C = \angle F$

$$\frac{x}{n} = \frac{y}{m} = \frac{z}{k}$$

Note:

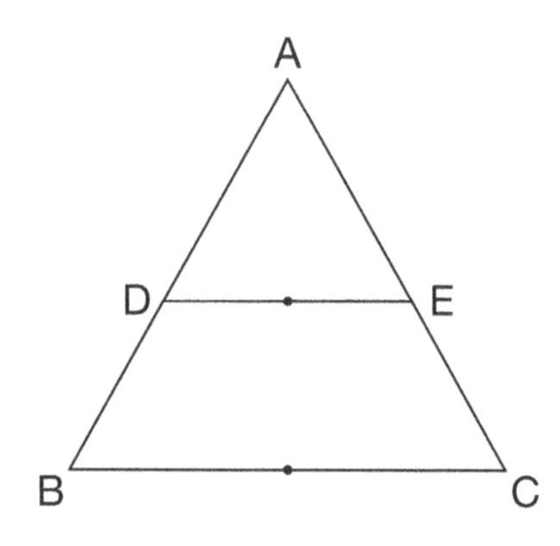

If [DE] // [BC], then

$$\frac{|AD|}{|AB|} = \frac{|AE|}{|AC|} \text{ or } \frac{|AD|}{|AE|} = \frac{|BD|}{|CE|}$$

Similarity Theorem Worksheet

Write a proportion and solve for x in each problem.

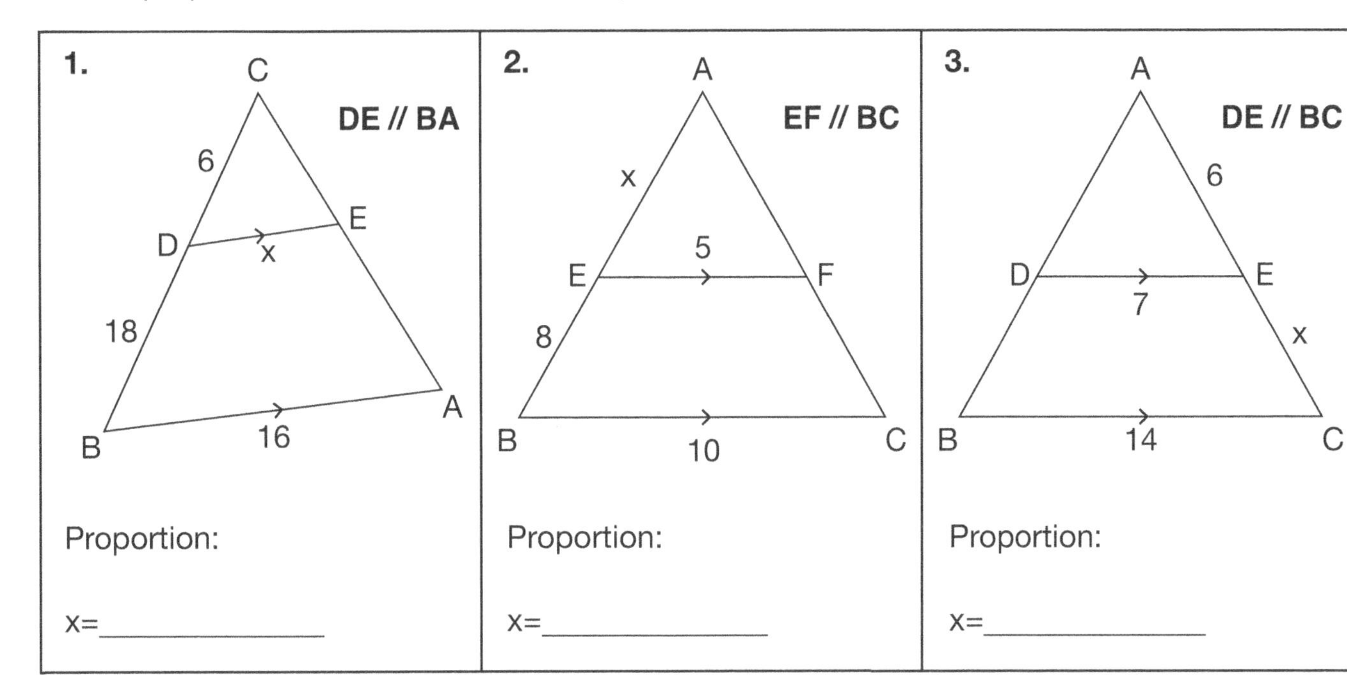

Proportion:

x=________________

Proportion:

x=________________

Proportion:

x=________________

Find the following missing length. The triangles in each pair are similar.

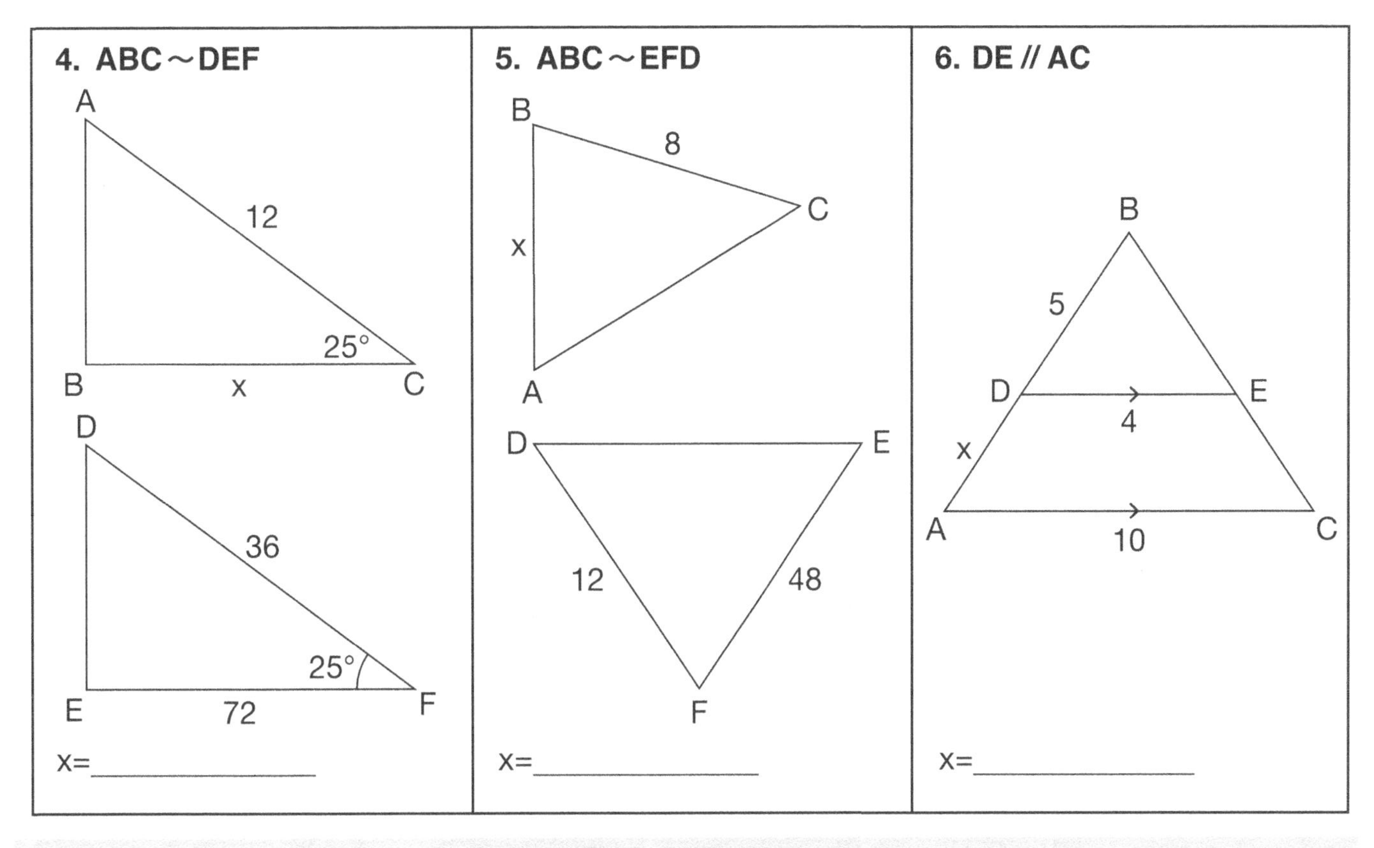

x=________________

x=________________

x=________________

Similarity Theorem Challenge

1.

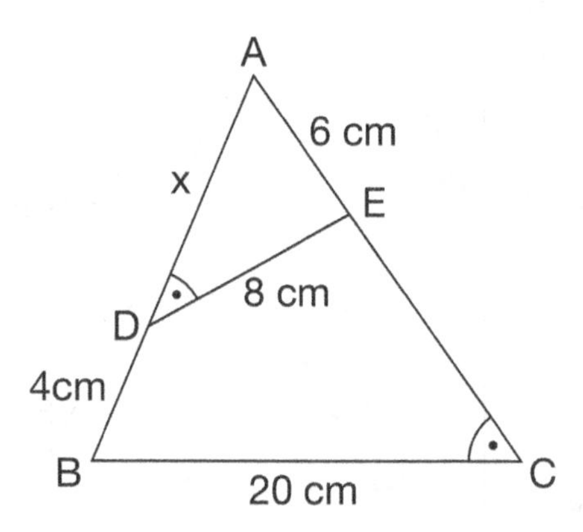

From the figure above, if $\angle D = \angle C$, and $|BC| = 20cm$, $|DE| = 8cm$, $|EA| = 6cm$, and $|BD| = 4cm$, then what is the value of x?

A) 5 cm B) 9 cm C) 11 cm D) 15 cm

2.

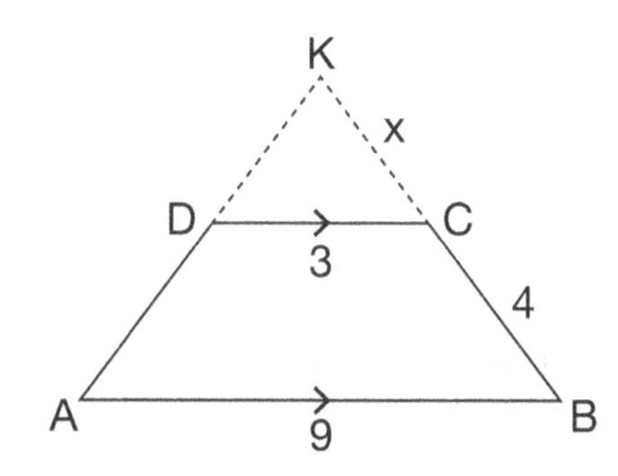

If DC // AB in the figure above, what is the value of x?

A) 2 B) 4 C) 6 D) 8

3.

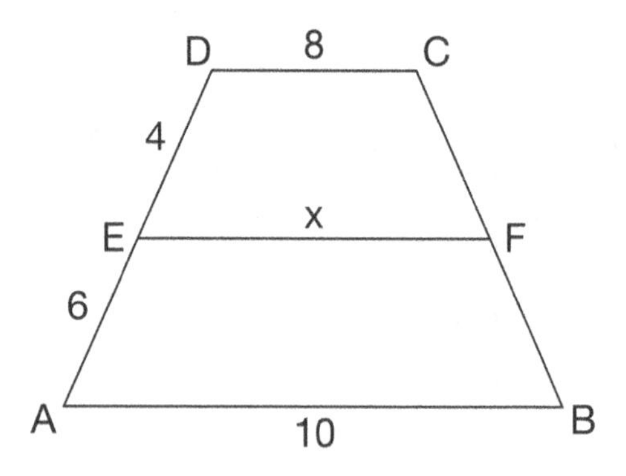

In the trapezoid ABCD above, DC = 8, EF = x, and AB = 10.

What is the value of x?

A) 2.2 B) 4.4 C) 8.8 D) 9.8

4.

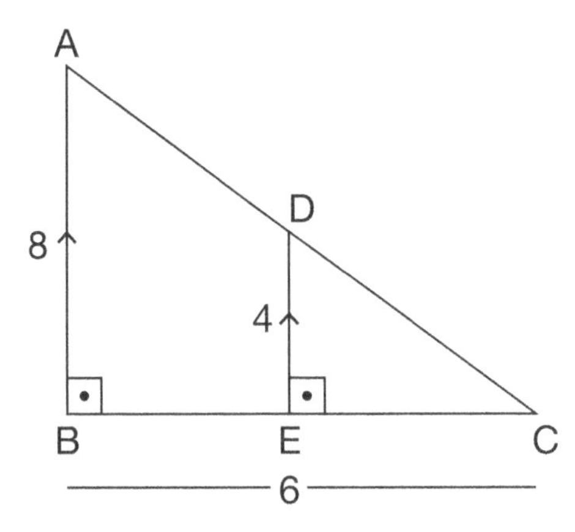

AB // DE

|AB| = 8

|BC| = 6

|DE| =4, then find |DC| + |EC|?

A) 5 B) 6 C) 7 D) 8

5.

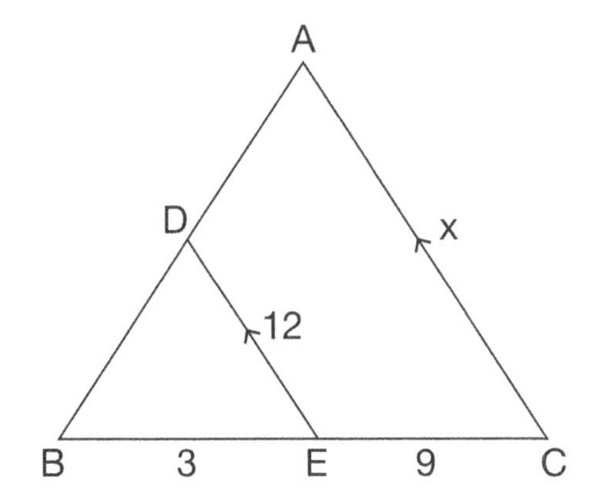

If DE // AC in the above figure, what is the value of x?

A) 12 B) 24 C) 36 D) 48

6.

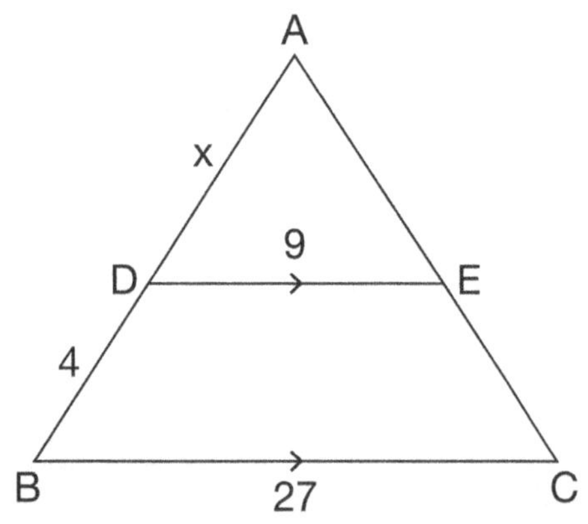

If DE // BC in the figure above, what is the value of x?

A) 2 B) 4 C) 6 D) 8

7.

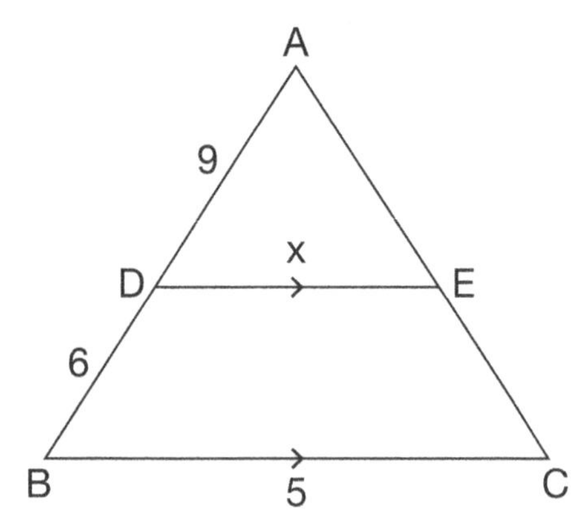

If DE // AC in the figure above, what is the value of x?

A) 3 B) 4 C) 5 D) 6

8.

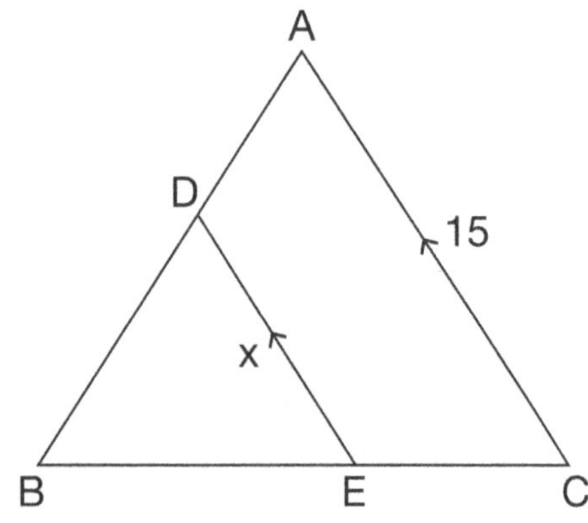

$3|AD| = 2|BD|$

If DE // AC in the figure above, what is the value of x?

A) 3 B) 6 C) 9 D) 12

9.

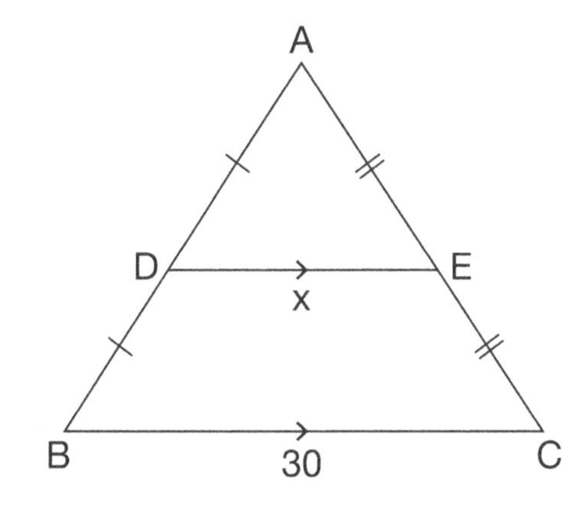

If DE // BC in the figure above, what is the value of x?

A) 5 B) 10 C) 15 D) 20

10.

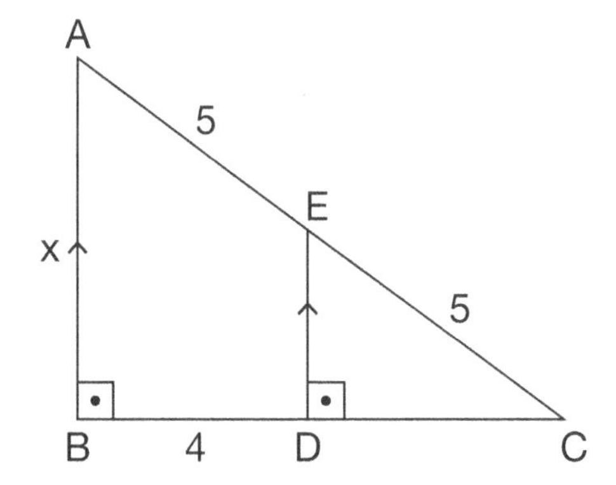

If AB // DE in the figure above, what is the value of x?

A) 3 B) 4 C) 6 D) 8

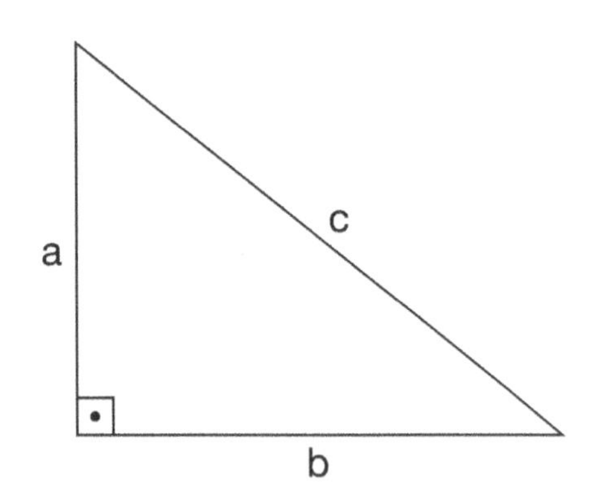

$a^2 + b^2 = c^2$

Example: What is the value of x?

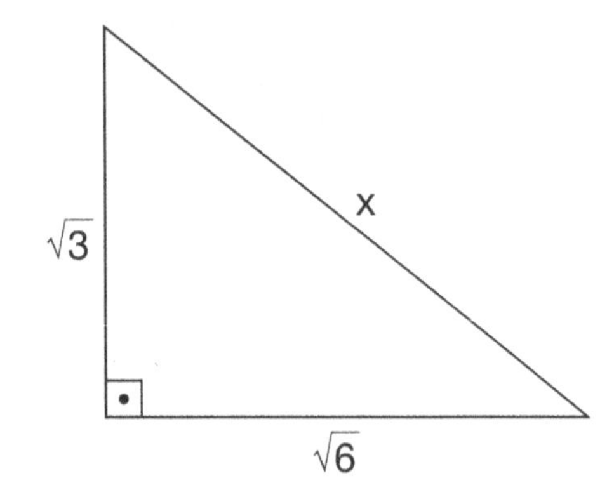

Solution:

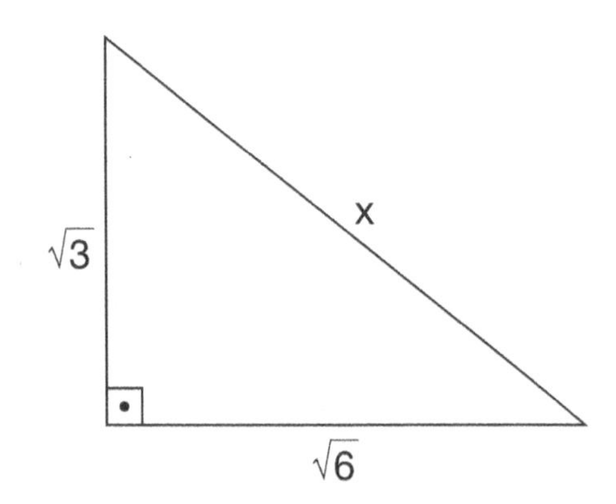

$(\sqrt{3})^2 + (\sqrt{6})^2 = x^2$

$3 + 6 = x^2$

$9 = x^2$

$3 = x$

Pythagorean Theorem Worksheet

Use the Pythagorean theorem to find distances on a coordinate grid

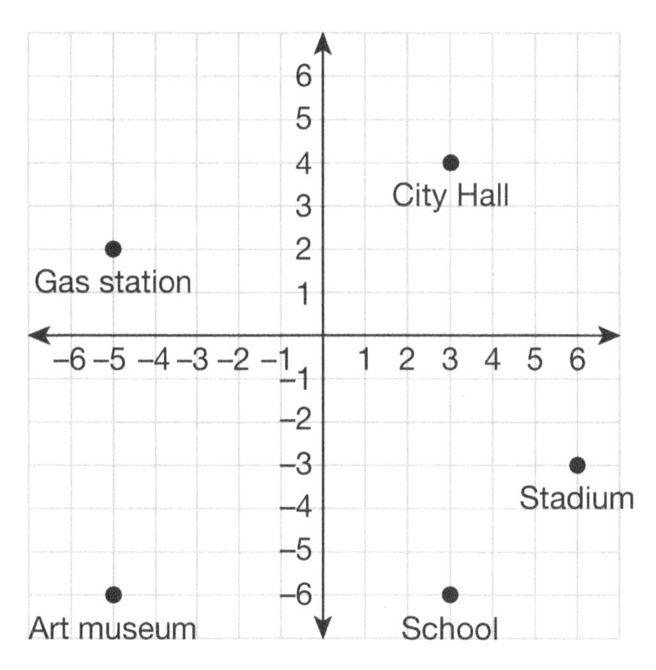

1. The gas station to city hall

2. The art museum to school

3. The city hall to stadium

4. The school to stadium

For each of the following triangles, find the missing length.

5.

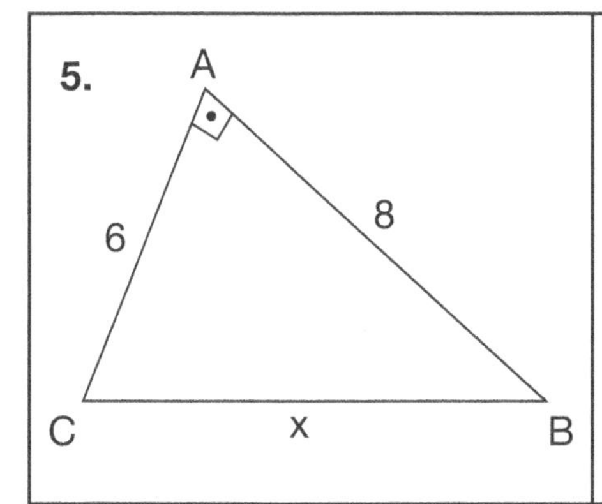

6.

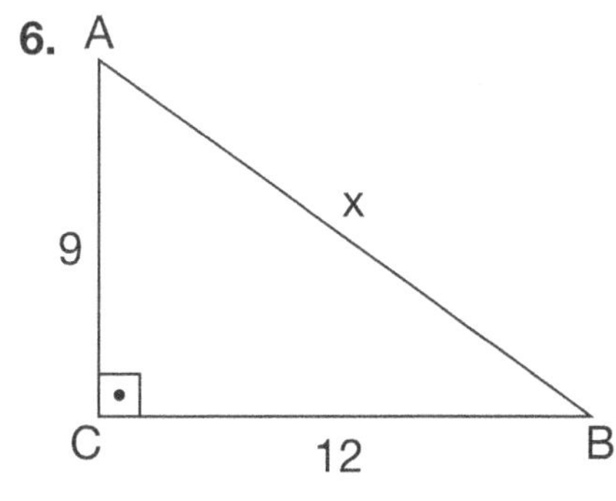

7.

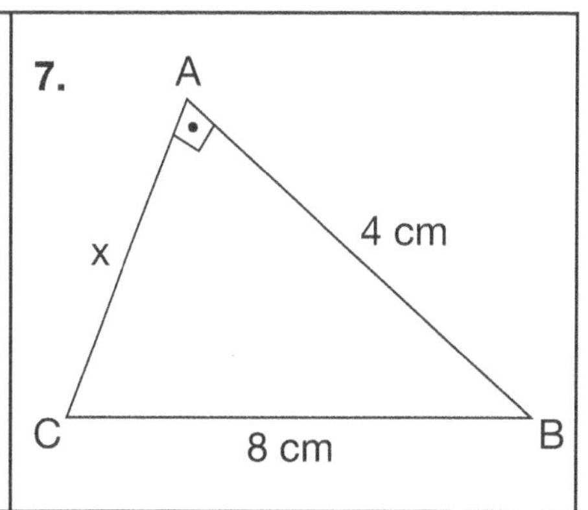

Pythagorean Theorem Challenge

1.

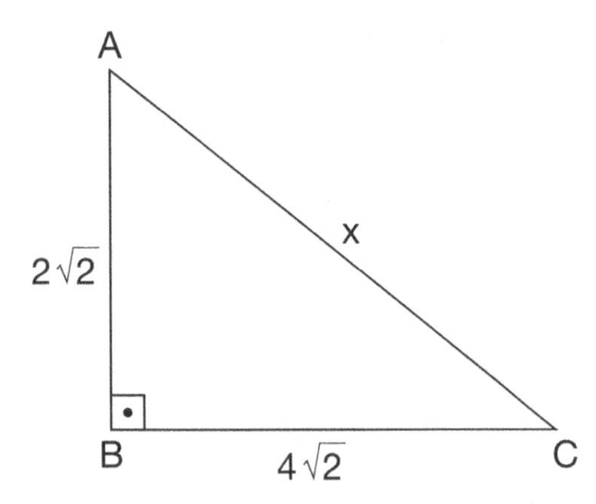

What is the value of x?

A) $\sqrt{10}$ B) $2\sqrt{10}$ C) $3\sqrt{10}$ D) 5

2.

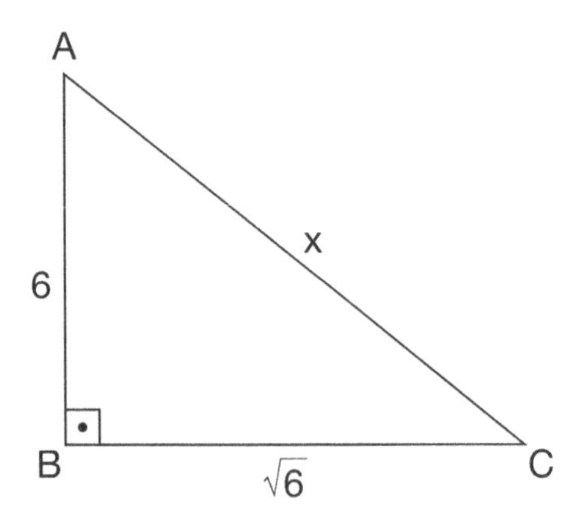

What is the value of x?

A) 42 B) $\sqrt{42}$ C) 46 D) $\sqrt{46}$

3.

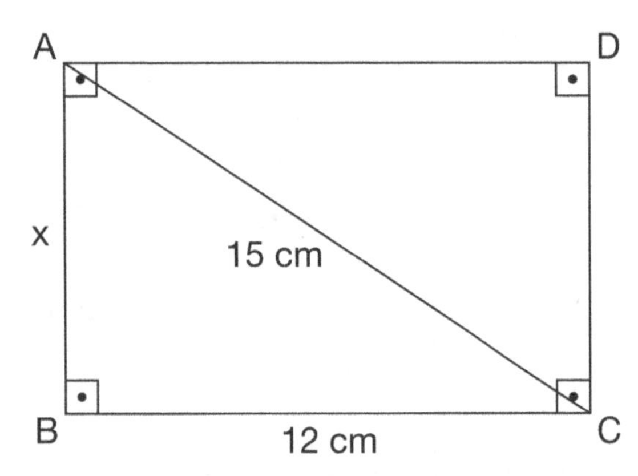

What is the value of x?

A) 3 cm B) 5 cm C) 9 cm D) 12 cm

4.

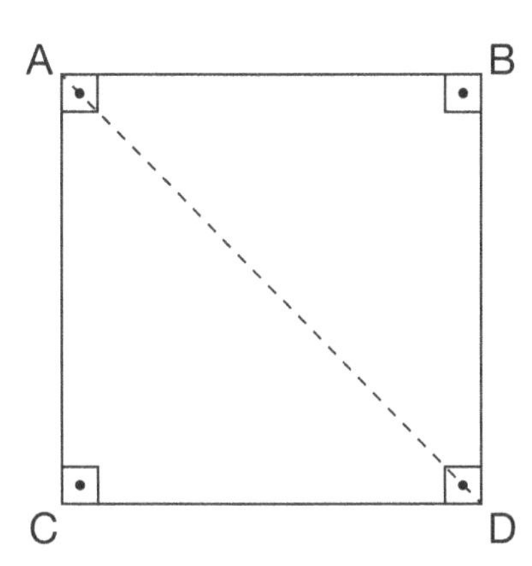

What is the area of the square above square if the length of AD is $6\sqrt{2}$?

A) 16 B) 25 C) 36 D) 47

5.

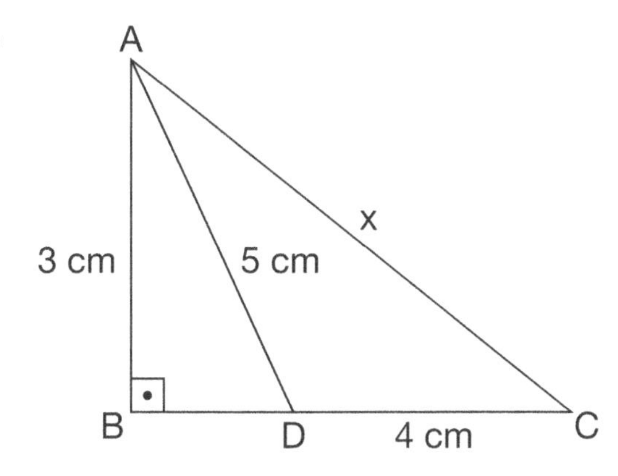

What is the value of x?

A) $\sqrt{73}$ cm B) $\sqrt{83}$ cm C) 8 cm D) 10 cm

6.

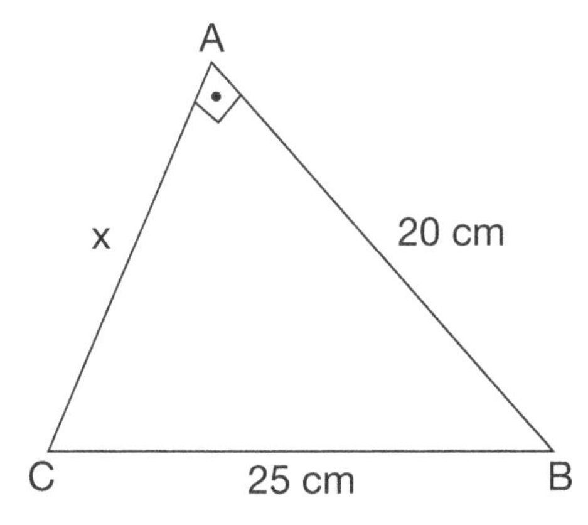

What is the value of x?

A) 9 cm B) 12 cm C) 14 cm D) 15 cm

7.

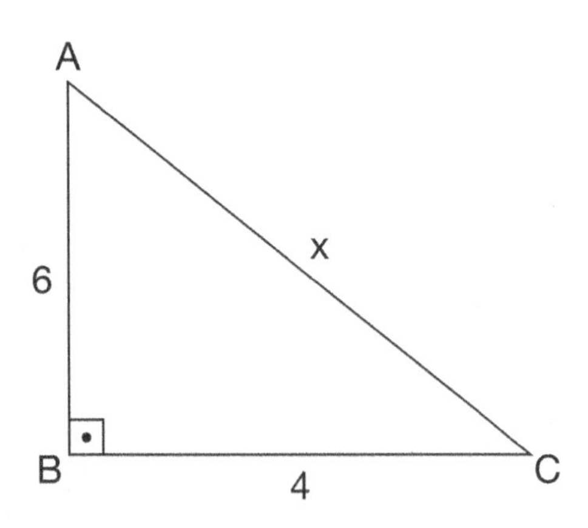

What is the value of x?

A) $\sqrt{13}$ B) $\sqrt{26}$ C) $2\sqrt{13}$ D) $2\sqrt{26}$

8.

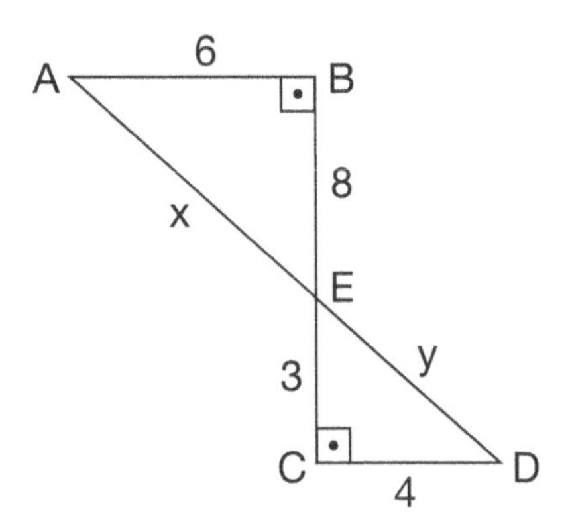

If AB // CD in the above figure, what is the value of x + y?

A) 5 B) 10 C) 12 D) 15

9.

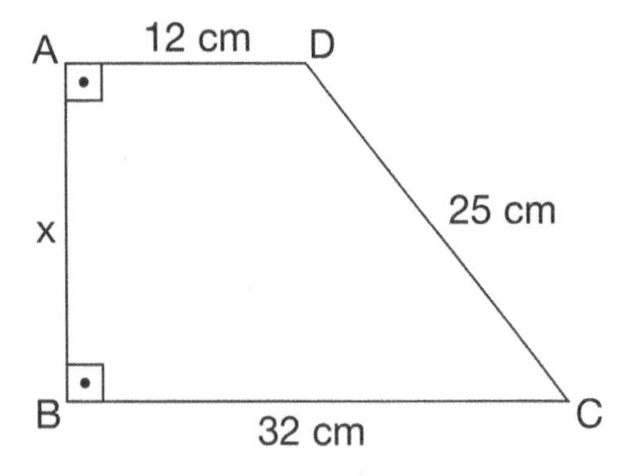

What is the value of x?

A) 10 cm B) 15 cm C) 20 cm D) 25 cm

COORDINATE PLANE

Coordinate Plane: A coordinate system formed by the intersection of a vertical number line, called the y–axis, and a horizontal number line, called the x–axis.

Origin: A beginning or starting point. The point where lines intersect each other at (0, 0).

Quadrant: One of four regions into which the coordinate plane is divided by the x- and y–axis. These regions are called the quadrants

Ordered pair: A pair of numbers that can be used to locate a point on a coordinate plane. The order of the numbers in a pair is important, and the x–axis always comes before the y–axis.

Ex: (3, 5)

x-coordinate: The first number in an ordered pair is called the x–coordinate.

Ex: (4,0)

y-coordinate: The second number in an ordered pair is called the y-coordinate.

Ex: (0,7)

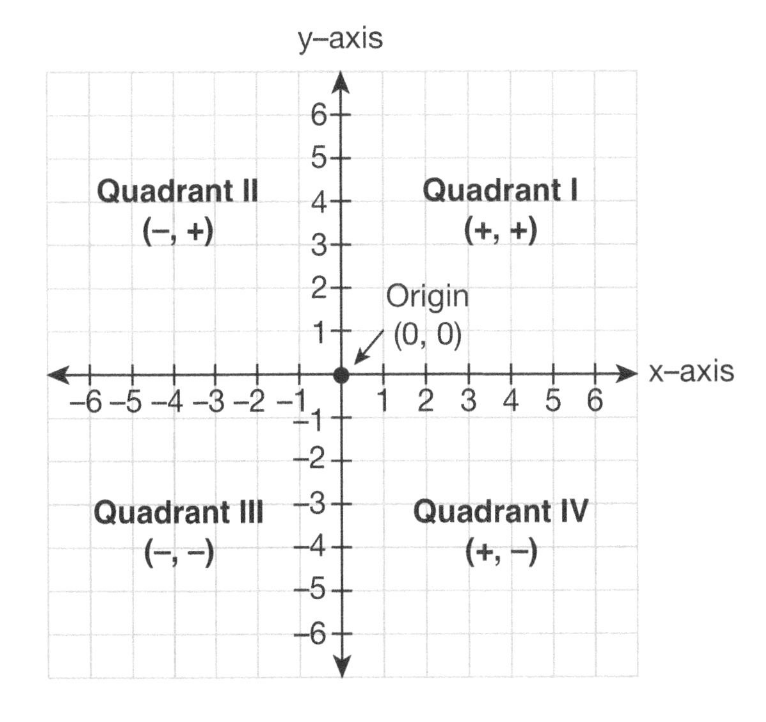

Coordinate Plane Worksheet

List the coordinates for each given point. Give the quadrant for each point.

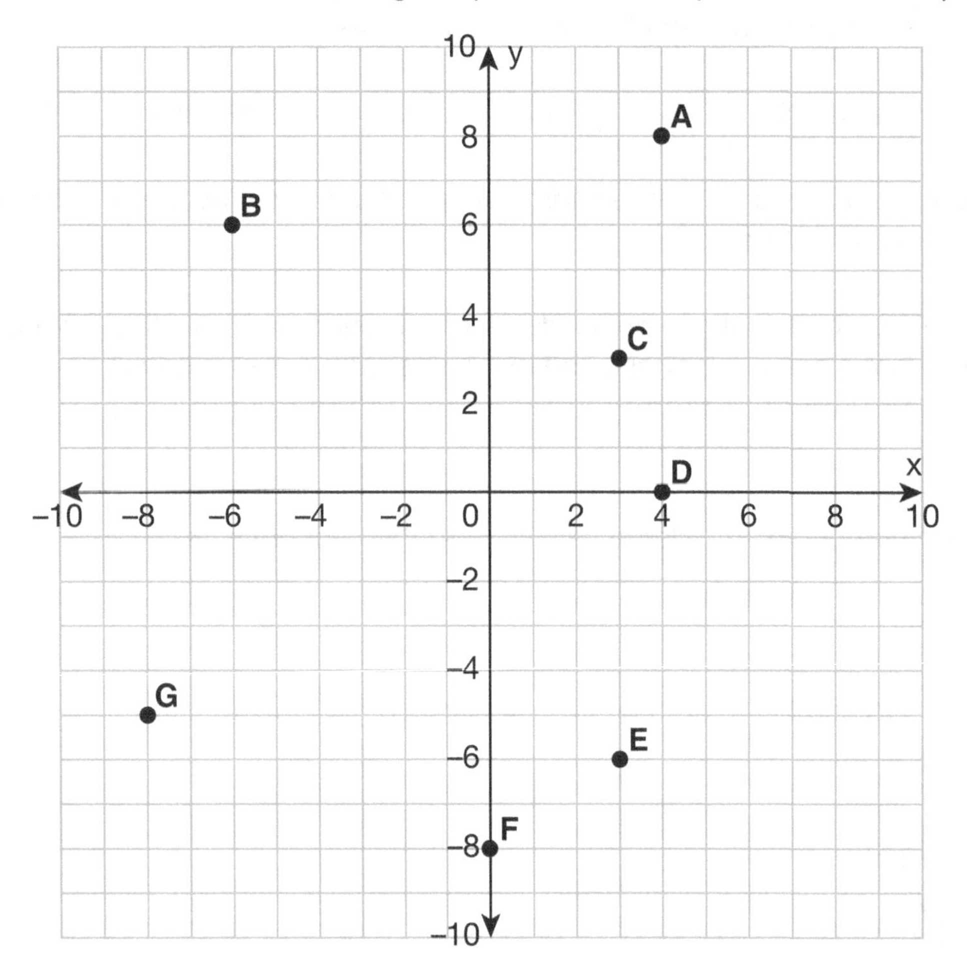

Coordinates	Point	Quadrant
	A	
	B	
	C	
	D	
	E	
	F	
	G	

Coordinate Plane Challenge

1. What are the coordinates of point A?

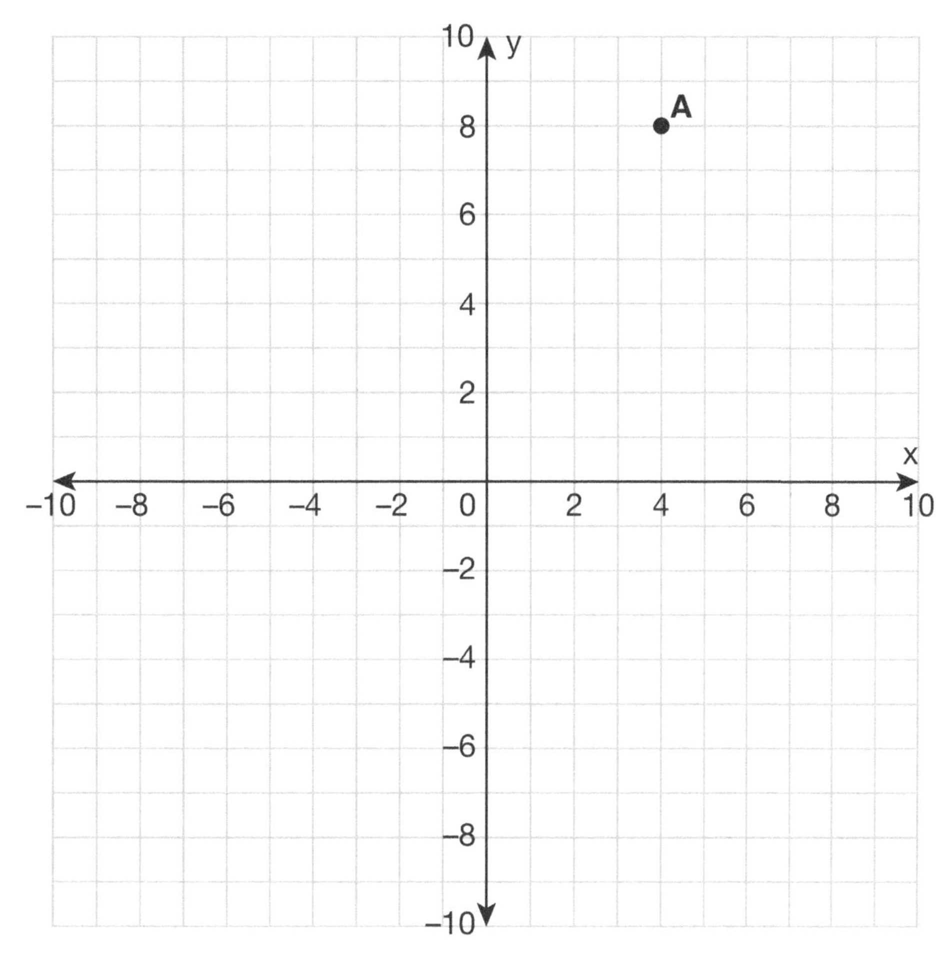

A) (8, 4) B) (4, 8) C) (0, 8) D) (4, 0)

2. Which pair of points has a distance of 10 units between them?

A) (0, 9) and (1, 9) B) (5, 9) and (1, 9) C) (15, 9) and (5, 9) D) (12, 9) and (1, 9)

3. Which ordered pair locates a point on the y–axis?

A) (1, 1) B) (1, 0) C) (3, 0) D) (0, −1)

Coordinate Plane Challenge

4. What are the coordinates of point C?

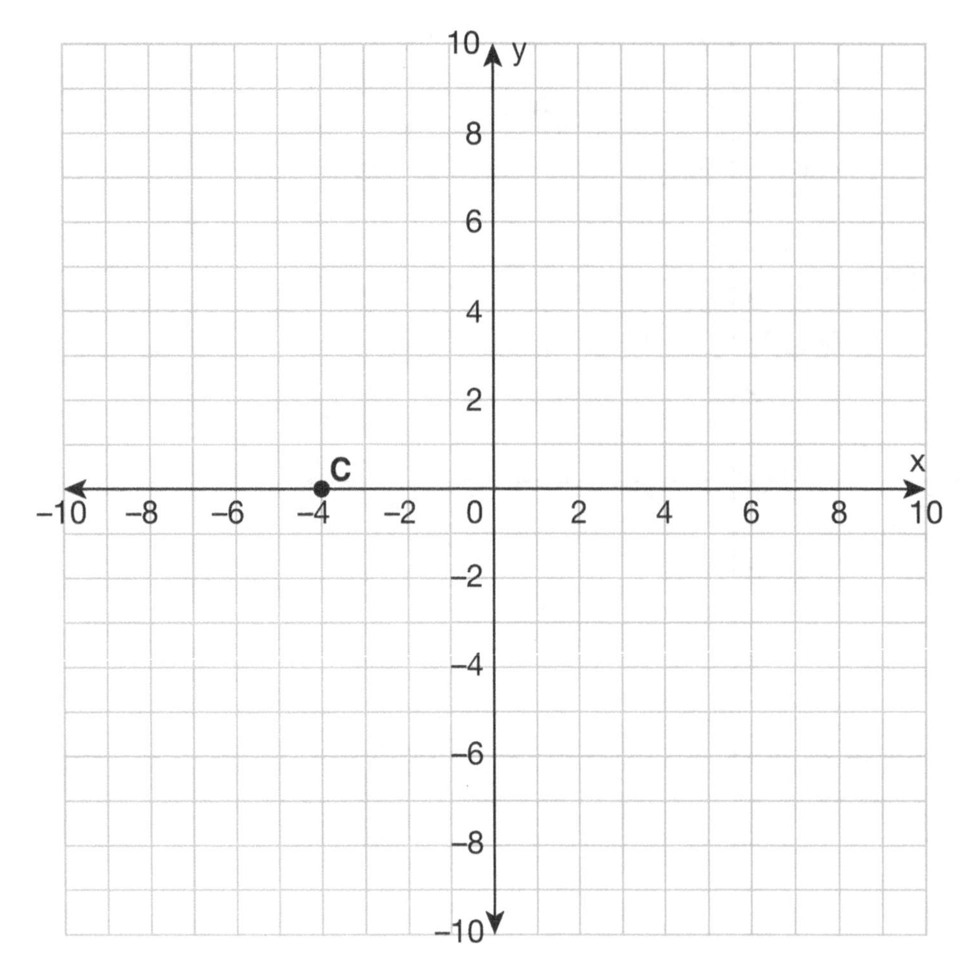

A) (0, 4) B) (0, 0) C) (0, −4) D) (−4, 0)

5. What is the distance from (4, 3) to the origin?

A) 2 units B) 3 units C) 4 units D) 5 units

6. The distance between (9, 16) and (3, y) is 10 units. Which of following can be the value of y?

A) 7 B) 12 C) 16 D) 24

7. Name the quadrant the point (−4,8) is in.

A) Quadrant I B) Quadrant II C) Quadrant III D) Quadrant IV

AREA & PERIMETER

Parallelogram	Area = base × height $P = 2(a + b)$	b, h, Height, Base
Triangle	$\text{Area} = \frac{\text{base} \times \text{height}}{2}$ $P = a + b + c$	a, c, h, b
Rectangle	Area = base × height $P = 2(h + b)$	b, h, Height, Base
Equilateral triangle	$\text{Area} = \frac{\sqrt{3}}{4} \times a^2$ $P = 3a$	a, a, a
Square	Area = width × length $P = 4s$	S, S, Area = s^2
Trapezoid	$\text{Area} = \frac{(b_1 + b_2)}{2} \times h$ $P = m + n + b_1 + b_2$	b_1, m, h, n, b_2

Area & Perimeter Worksheet

Find the length of the unknown sides given the perimeters of the following figures.

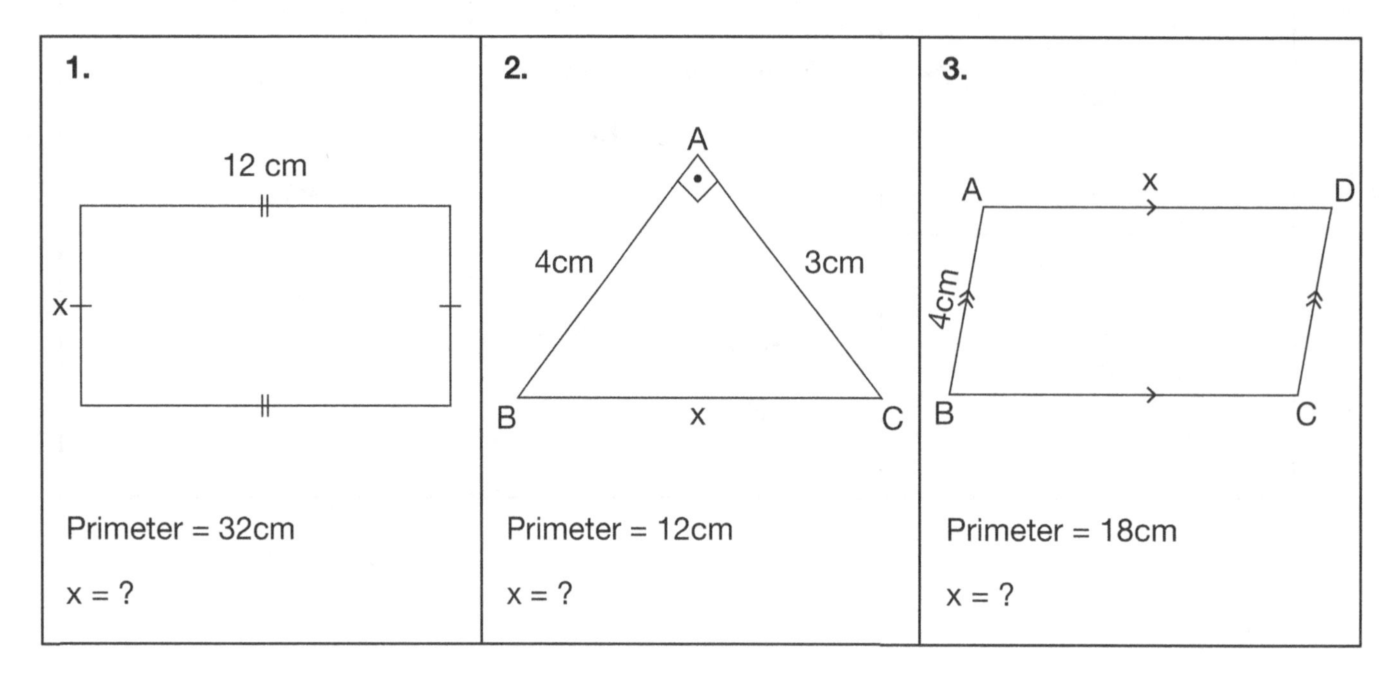

Find the area of the following figures.

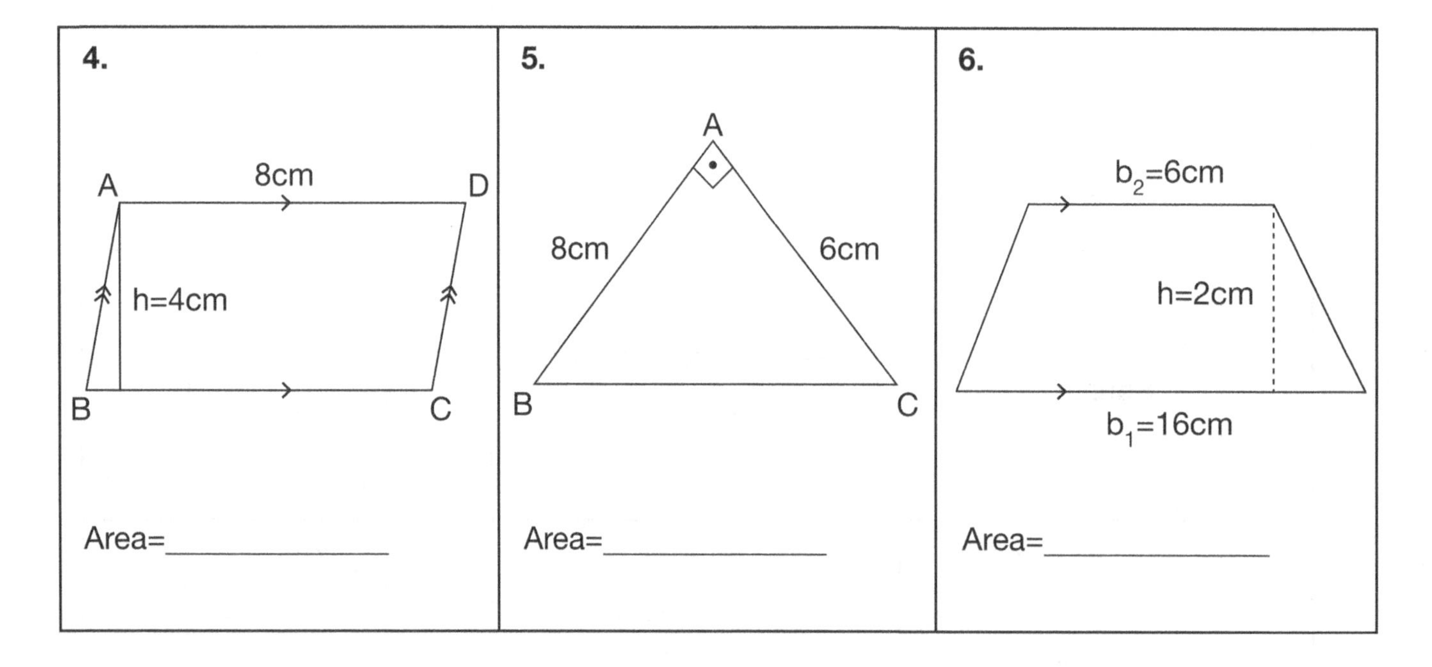

Area & Perimeter Challenge

1. Find the perimeter of the rectangle.

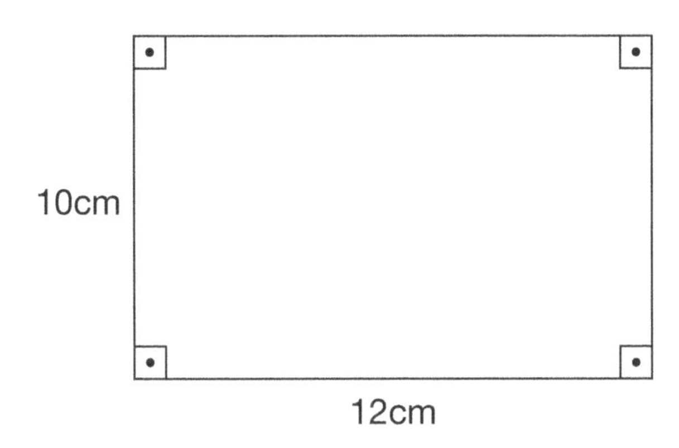

A) 22cm B) 32cm C) 44cm D) 54cm

2. Find the area. (The figure is not drawn to scale)

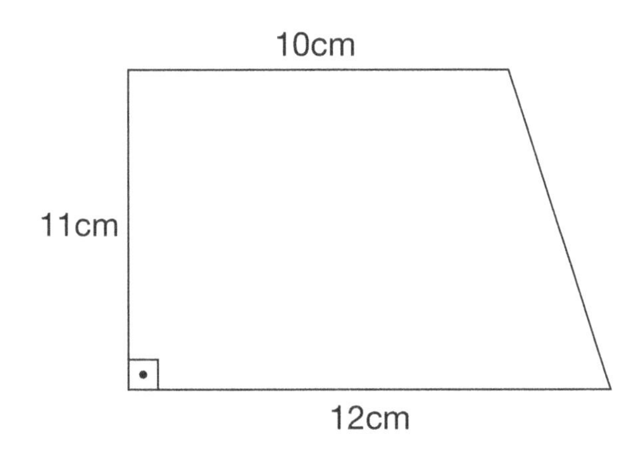

A) $22cm^2$ B) $44cm^2$ C) $108cm^2$ D) $121cm^2$

3. Find the perimeter of the following shape.

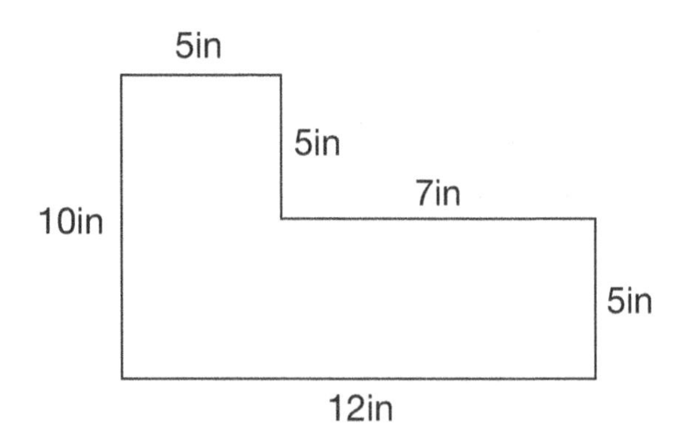

A) 17in B) 27in C) 44in D) 47in

4. Find the perimeter of the following triangle

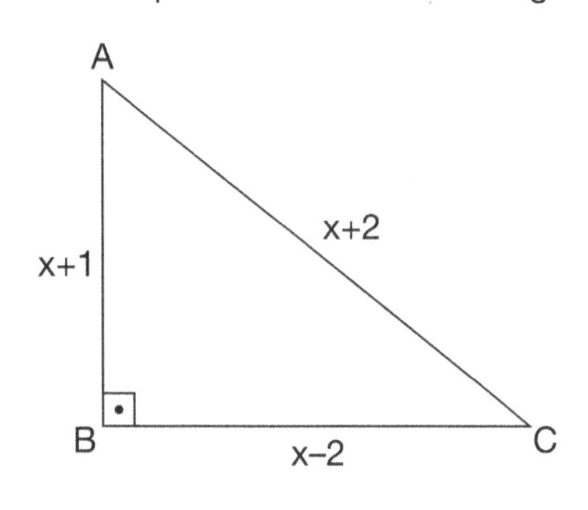

A) $3x$ B) $3x - 1$ C) $3x + 1$ D) $3x + 2$

5. Find the area of $\triangle$ ABC.

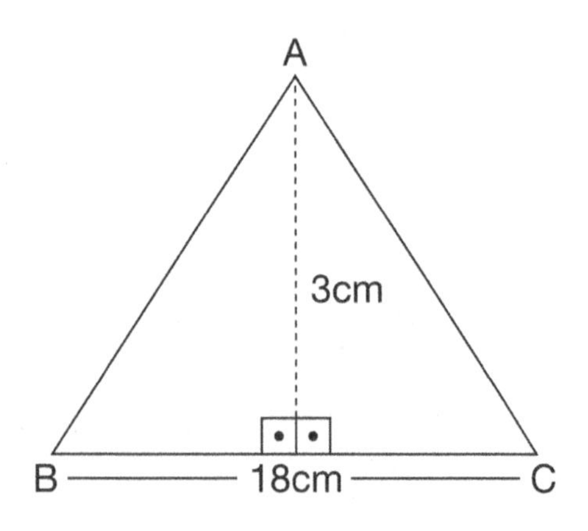

A) $18cm^2$ B) $27cm^2$ C) $30cm^2$ D) $36cm^2$

6. What is the area of Δ ABC?

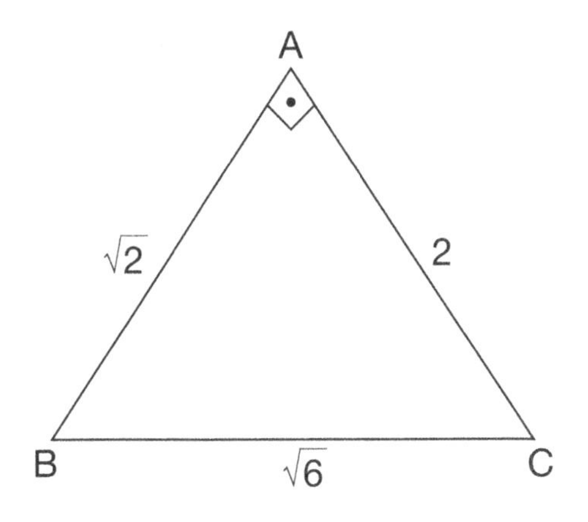

A) 2 B) $\sqrt{2}$ C) $2\sqrt{2}$ D) $3\sqrt{2}$

7. What is the area of the figures Δ ABC?

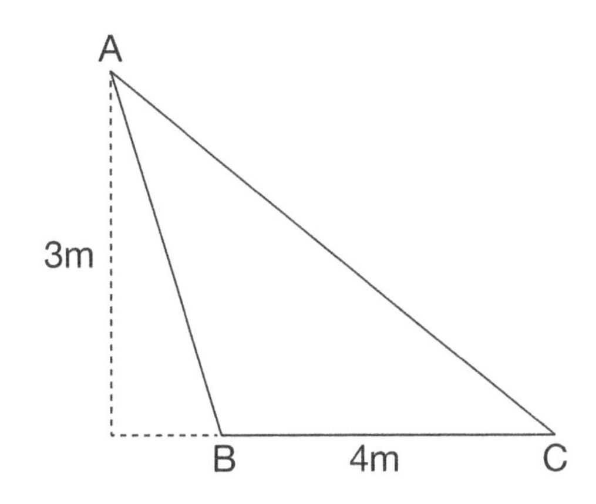

A) $3m^2$ B) $6m^2$ C) $9m^2$ D) $12m^2$

Area & Perimeter Challenge

8. What is the area of an equilateral triangle with a side of 4?

A) $\sqrt{3}$ B) $2\sqrt{3}$ C) $4\sqrt{3}$ D) $6\sqrt{3}$

9. What is the area of the following square if the length of AD is $6\sqrt{2}$?

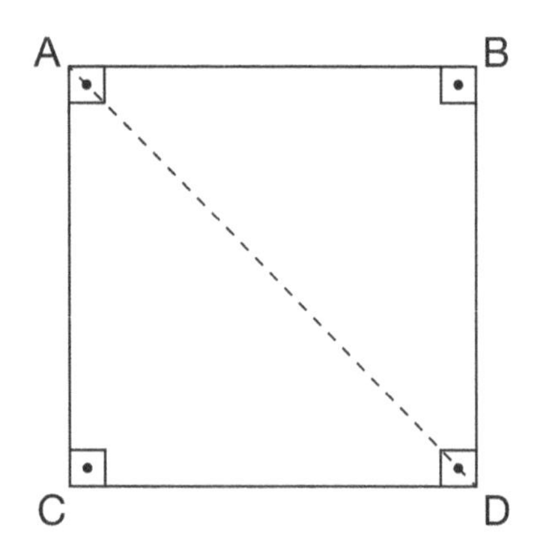

A) 16 B) 25 C) 36 D) 47

10. The perimeter of a rectangle is 48m. If the width of the rectangle is three times the length, what is the width?

A) 6cm B) 9cm C) 15cm D) 18cm

CIRCLES

Radius: The distance from the center of the circle to the edge.

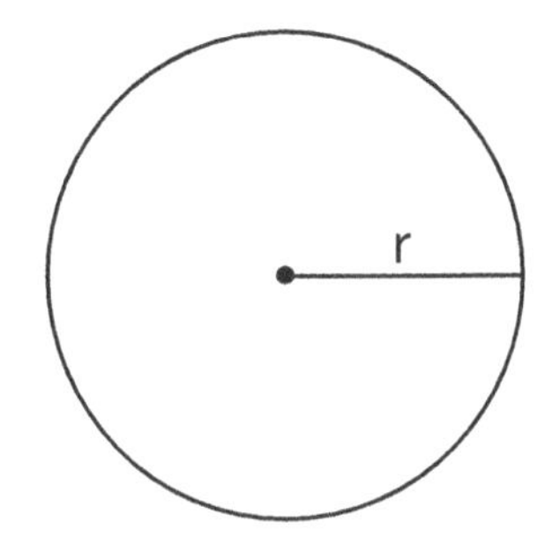

Chord: A line segment whose endpoints are on the circle.

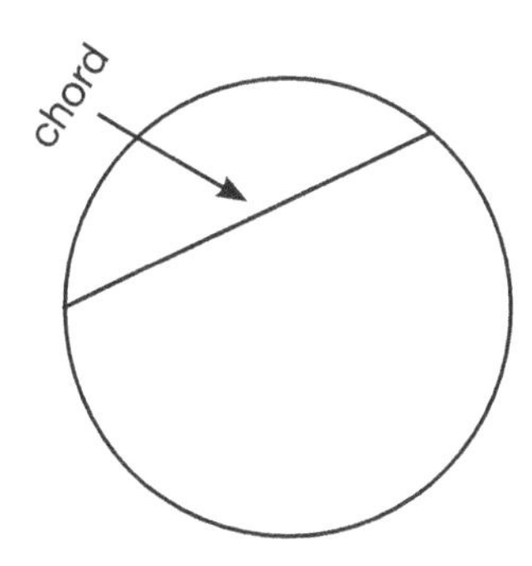

Diameter: A **chord** that passes through the center of the circle.

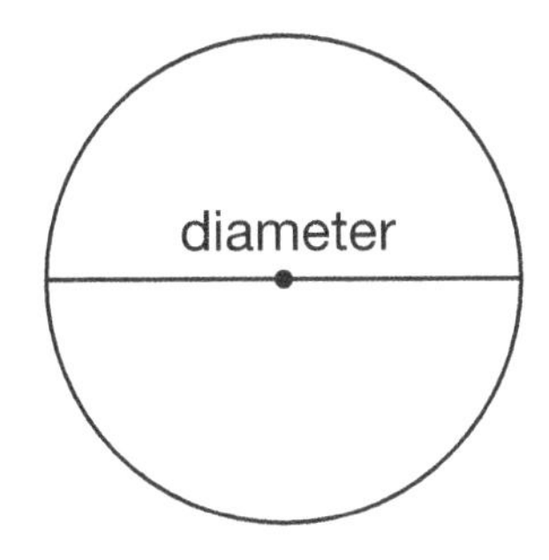

Tangent: A line which intersects a circle at exactly one point.

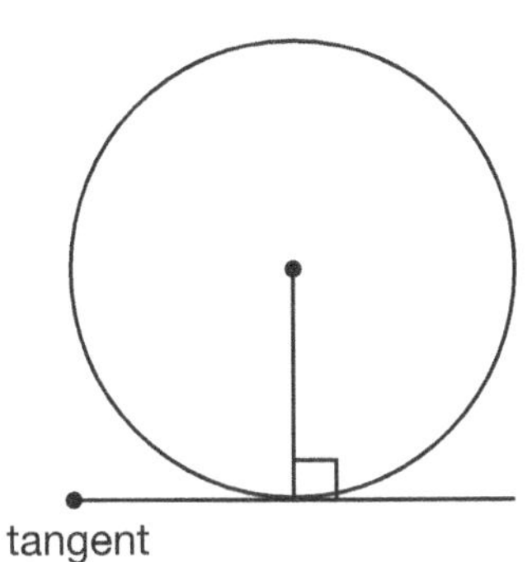

Segment: A region of a circle which is "cut off" from the rest of the circle by a secant or a chord.

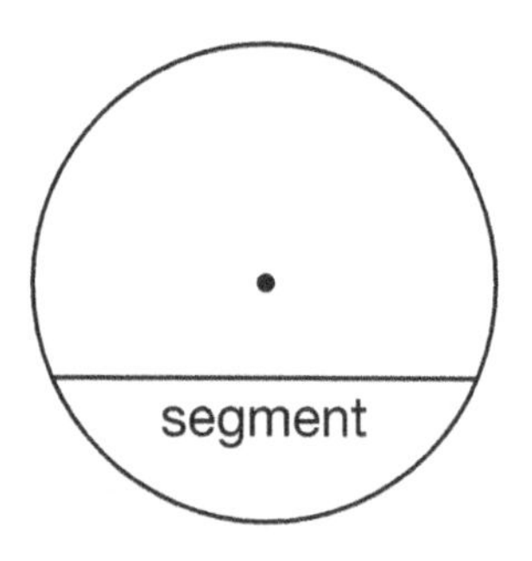

Area of a Circle:

$A = \pi r^2$

Circumference:

$C = 2\pi r$

Semi-Circle:

$A = \frac{1}{2}\pi r^2$

Circles Worksheet

Find the diameter, circumference, and area for each circle below. To find the area, leave your answer in terms of π

1.

r=5cm

Diameter:

Circumference:

Area:

2.

r=6ft

Diameter:

Circumference:

Area:

3.

r=10cm

Diameter:

Circumference:

Area:

Find the radius, circumference and the area for each circle below. To find the area, leave your answer in terms of π

4.

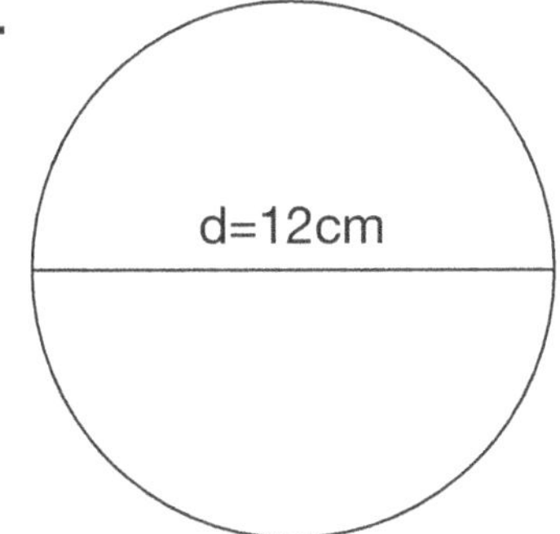

Radius:

Circumference:

Area:

5.

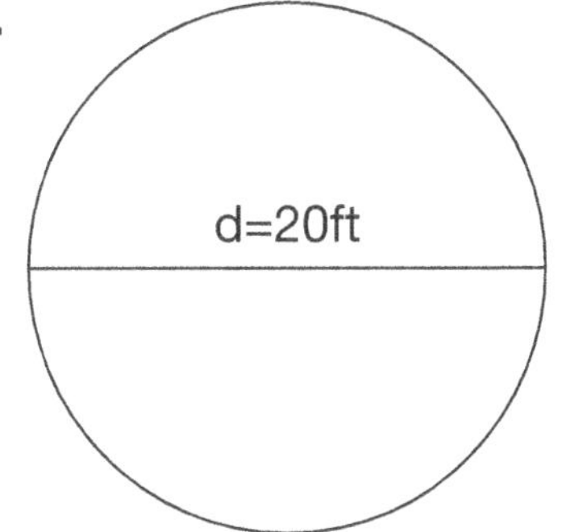

Radius:

Circumference:

Area:

6.

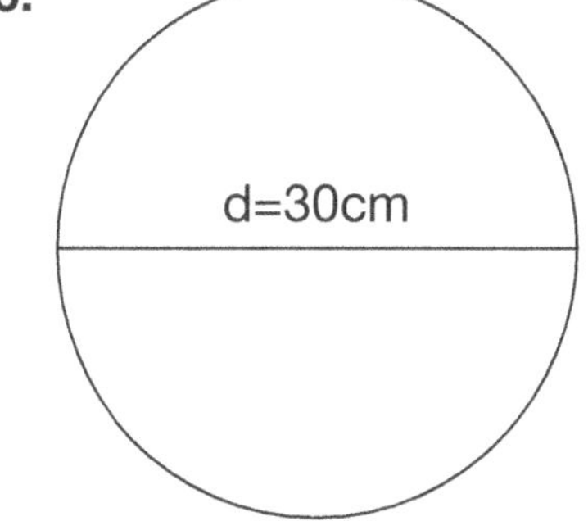

Radius:

Circumference:

Area:

Circles Challenge

1. Find the circumference of the following circle.

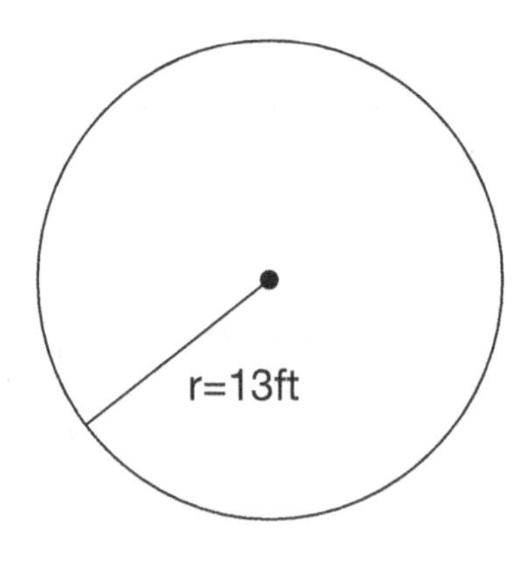

A) 13πft B) 26πft C) 36πft D) 42πft

2. Find the area of the following circle.

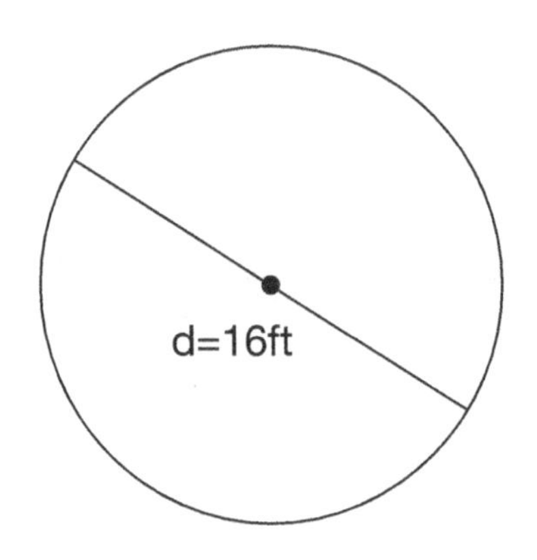

A) $16\pi ft^2$ B) $32\pi ft^2$ C) $48\pi ft^2$ D) $64\pi ft^2$

Circles Challenge

3. Using the two circles shown below, what is $\frac{\text{area of circle A}}{\text{area of circle B}}$?

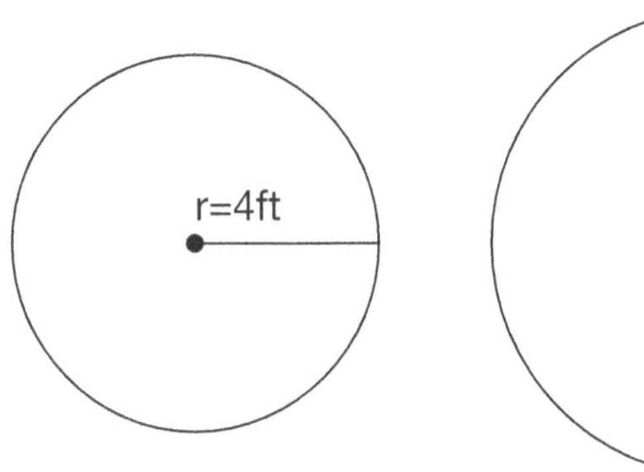

Circle A

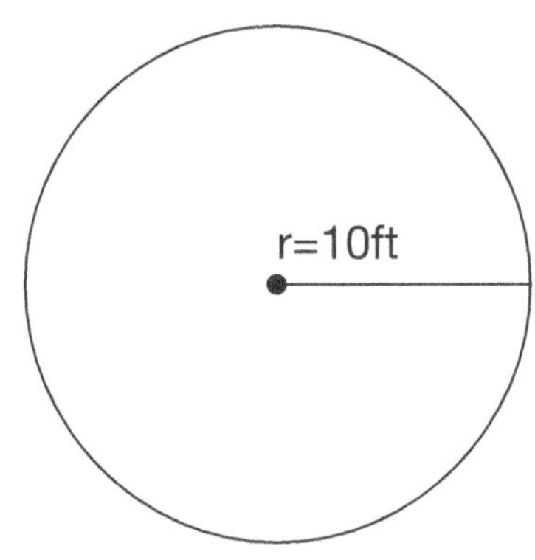

Circle B

A) $\frac{4}{25}$ B) $\frac{1}{5}$ C) $\frac{3}{4}$ D) $\frac{4}{5}$

4. In the following figure, O is the center of the circle and the radius is 4cm. Find the area of the shaded part.

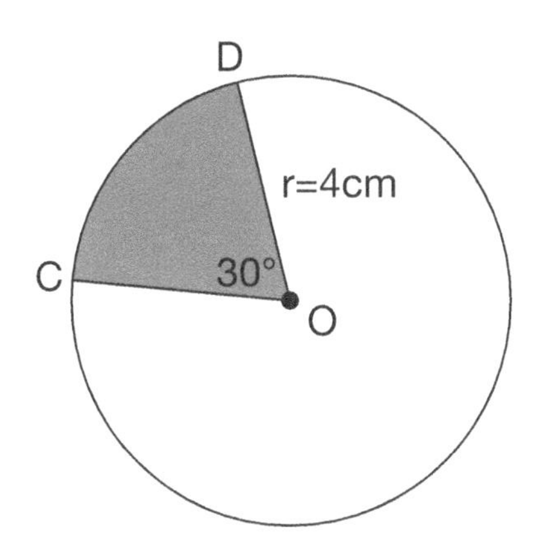

A) $\frac{3}{4}\pi cm^2$ B) $\frac{4}{3}\pi cm^2$ C) $4\pi cm^2$ D) $16\pi cm^2$

5. If the area of a circle is 16π, then find the circumference of the circle.

A) 4π B) 8π C) 10π D) 12π

Circles Challenge

6. If the circumference of a circle is 12π, what is the area of the circle? (Give your answer in terms of π.)

A) 6π B) 12π C) 24π D) 36π

7. If the ratio of the circumference to the area of the circle is 4 to 6, what is the radius of the circle?

A) 1   B) 3   C) 4   D) 6

8. Using the two circles shown below, what is $\frac{\text{circumference of circle X}}{\text{circumference of circle Y}}$?

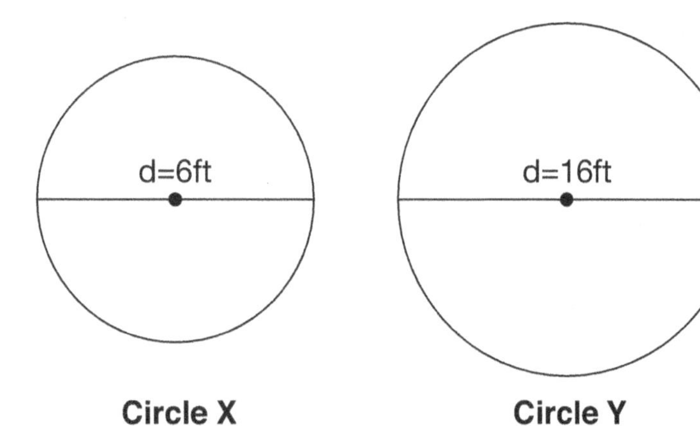

A) $\frac{3}{8}$ B) $\frac{3}{4}$ C) $\frac{3}{2}$ D) 3

VOLUME

Right rectangular prism

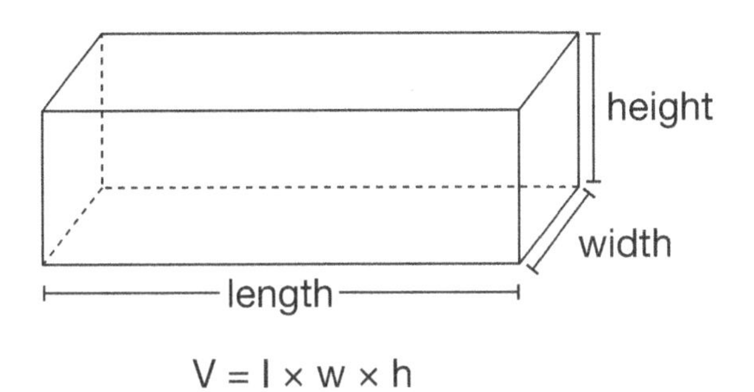

Cube

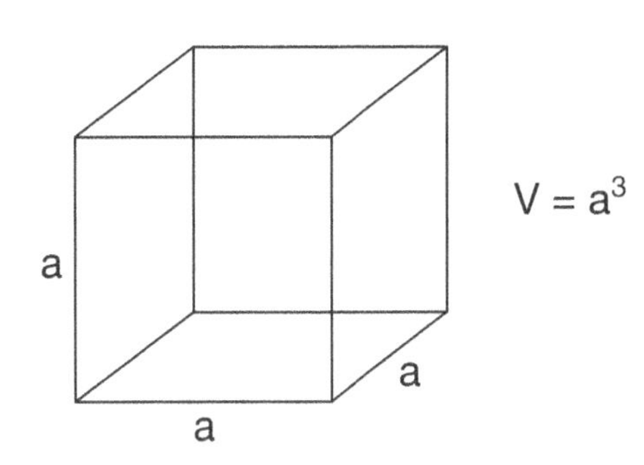

Cylinder

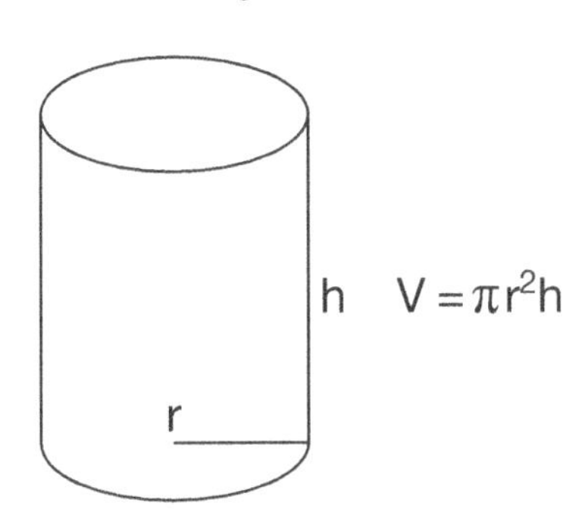

Cone

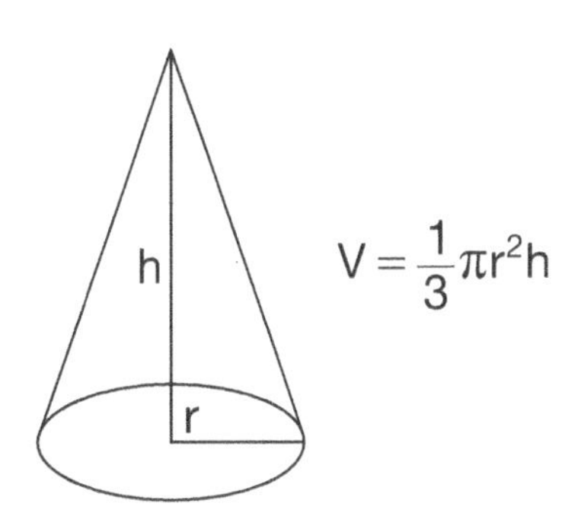

Pyramid

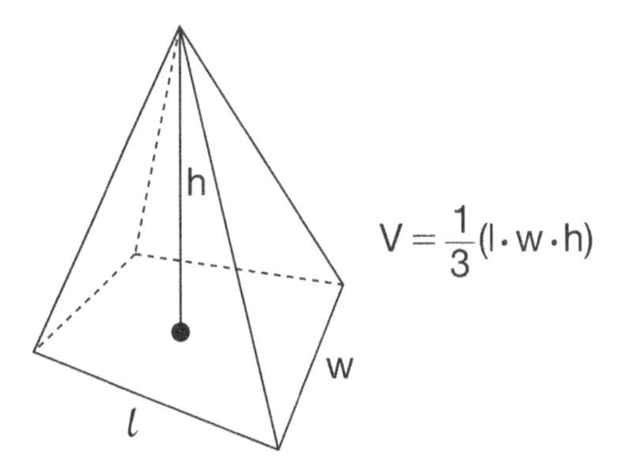

Sphere

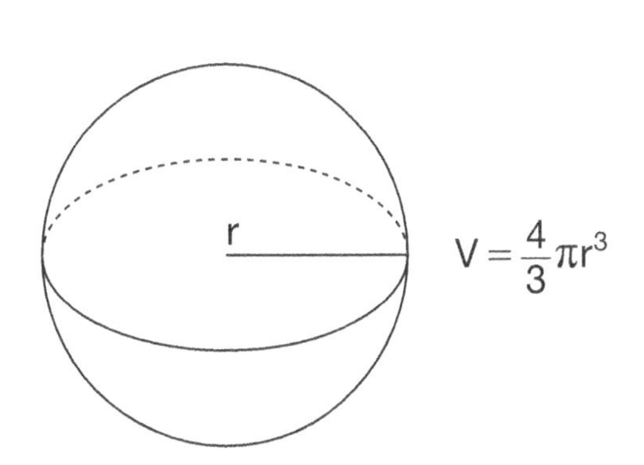

Volume Worksheet

Find the volume of each cube

1.

7ft

7ft

7ft

Volume: ______________

2.

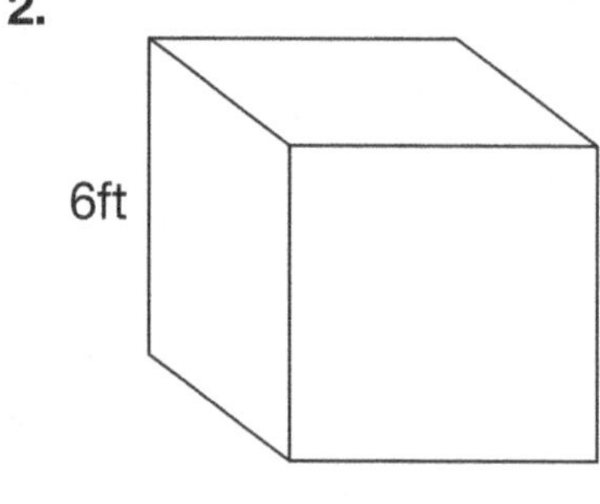

Volume: ______________

3.

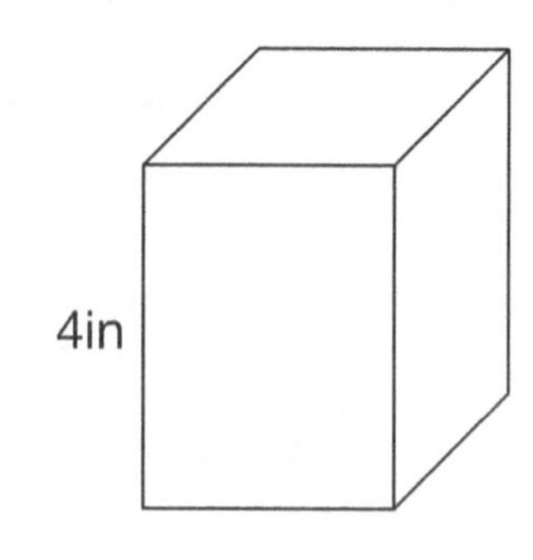

Volume: ______________

Find the volume of each rectangular prism

4.

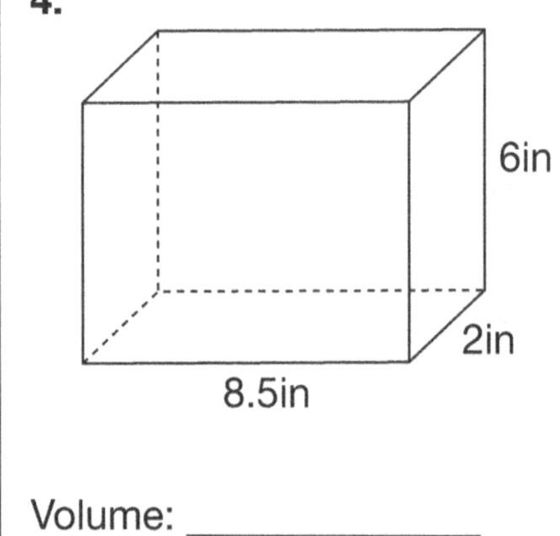

Volume: ______________

5.

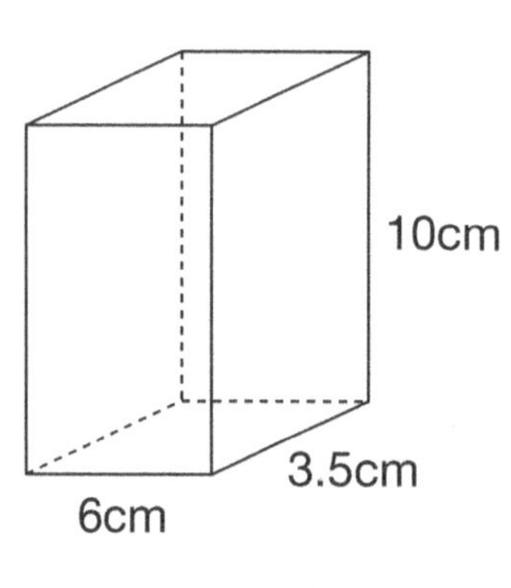

Volume: ______________

6.

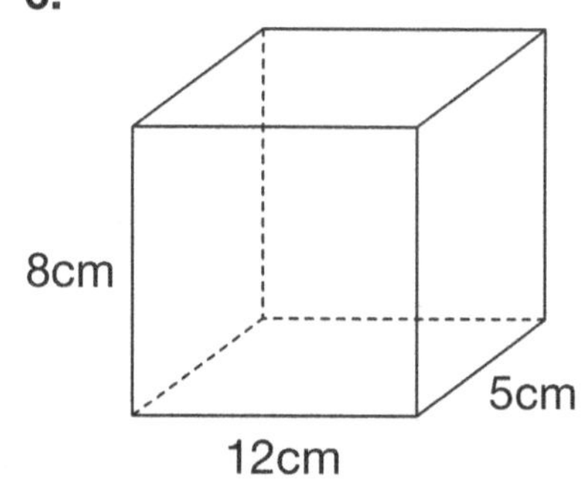

Volume: ______________

Volume Challenge

1. Find the volume of the following cube.

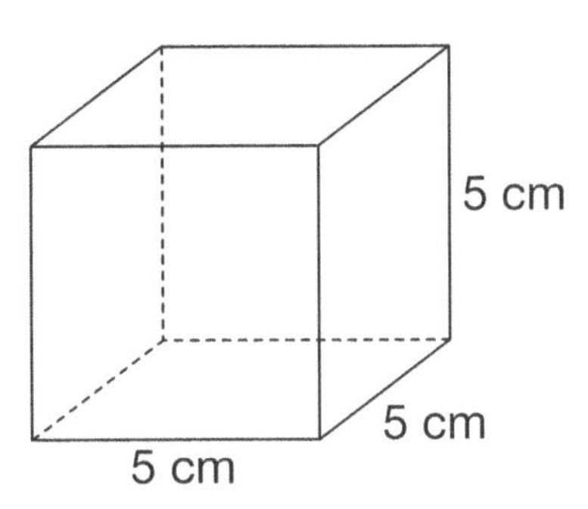

A) $25cm^3$ B) $125cm^3$ C) $215cm^3$ D) $325cm^3$

2. A cube has a volume of 27 ft^3. What is the length of one side of the cube?

A) 1ft B) 2ft C) 3ft D) 4ft

3. The volume of the following rectangular prism is 96 cm^3. What is value of h?

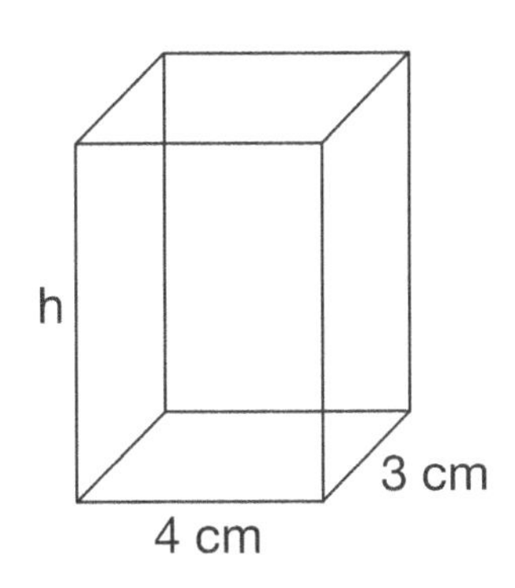

A) 4cm B) 8cm C) 12cm D) 16cm

4. Find the volume of a rectangular prism that has a length of 2cm, a width of 3cm, and a height of 6cm.

A) $16cm^3$ B) $18cm^3$ C) $30cm^3$ D) $36cm^3$

5. If the volume of the following cylinder is 72π cm^3, find the radius of the cylinder.

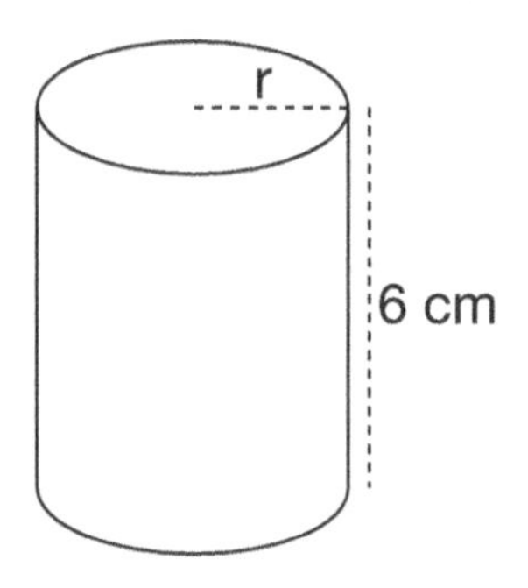

A) $\sqrt{3}$ cm   B) $2\sqrt{3}$ cm   C) $3\sqrt{3}$ cm   D) 6cm

6. Find the volume of the sphere shown below. (Give your answer in terms of π)

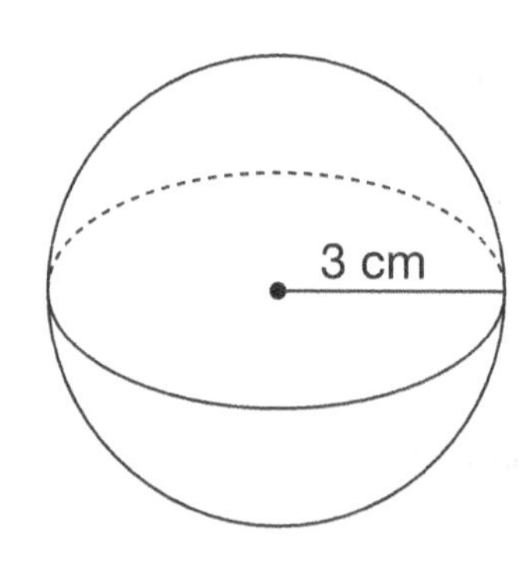

A) 10π cm^3 B) 18π cm^3 C) 36π cm^3 D) 49π cm^3

7. The volume of a sphere is 36π cm^3. Find the radius of the sphere.

A) 1cm   B) 3cm   C) 6cm   D) 8cm

8. Find the volume of the cylinder shown below. (Give your answer in terms of π)

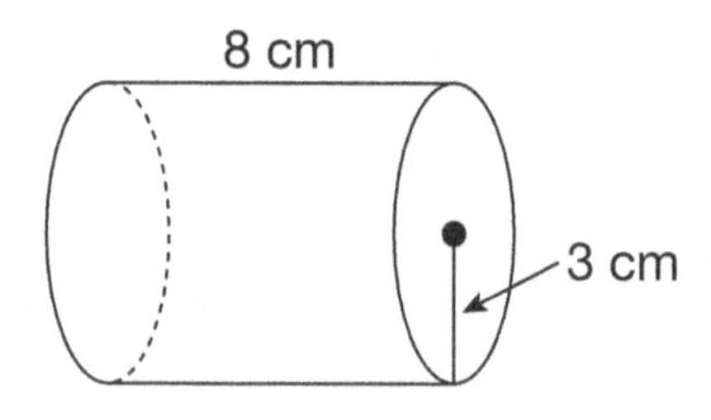

A) 36π cm^3 B) 72π cm^3 C) 84π cm^3 D) 96π cm^3

MEAN, MEDIAN, MODE & RANGE

Mean (Average): The sum of the numbers divided by how many numbers there are.

$$\text{Mean} = \frac{\text{Sum of the numbers}}{\text{How many numbers}}$$

Example: Find the mean of 3, 5, 6, 7, 9.

Solution:

$$\text{Mean} = \frac{\text{Sum of the numbers}}{\text{How many numbers}}$$

$$\text{Mean} = \frac{3+5+6+7+9}{\text{How many numbers}} = \frac{30}{5} = 6$$

Median: The middle number.

- ✓ Put all of the numbers from smallest to largest
- ✓ The median is the middle number.
- ✓ If there is an evan amount of numbers, the median is the average of the two middle numbers.

Example: Find the median of 3, 12, 20, 8, 16, 30.

- ✓ Solution: put all of the numbers into order. 3, 8, 12, 16, 20, 30 the middle numbers are 12 and 16. The average of these two number is $\frac{12+16}{2} = 14$

Mode: The most frequent value.

Example: Find the mode: A, B, C, D,C, A, B, A, C, D, C

Solution: The most repeated letter is C

Range: The difference between the lowest and highest value.

Example: Find the range of 2, 4, 20, 30, 50, 80.

Solution: Range = 80 − 2 = 78

Mean, Median, Mode, & Range Worksheet

Find the mean, median, mode, and range for each set of number.

1. 3, 4, 5, 6, 6, 7, 11

Mean:

Median:

Mode:

Range:

2. 1, 3, 5, 7, 7, 7

Mean:

Median:

Mode:

Range:

3. $\frac{1}{2}, \frac{1}{3}, \frac{1}{4}, \frac{1}{5}$

Mean:

Median:

Mode:

Range:

Find the missing number using the data points and the mean.

4. The average of 6 and y is 10.

y: ______________

5. The average of 12 and z is 20.

z: ______________

6. The average of 12.5 and k is 30.5.

k: ______________

7. The average of five consecutive positive integers is 30. What is the greatest possible value of one of these integers?

8. What measure of central tendency is calculated as the difference between the lowest and highest values?

Mean, Median, Mode, & Range Challenge

1. What is the mean of the following numbers?

12, 18, 24, 36, 20

A) 11 B) 13 C) 15 D) 22

2. The mean of five numbers is 48. If four of the numbers are 34, 54, 64 and 72 what is the value of the fifth number?

A) 4 B) 8 C) 12 D) 16

Use the data to answer question 3 and 4.

The temperature in Celsius (C°) in the first week of October was as follows:

25°, 21°, 20°, 25°, 23°, 14°, 32°

3. What is the mode of the temperatures for the first week of October?

A) 20°C B) 21°C C) 23°C D) 25°C

4. From the above data what is the median of the temperatures for the first week of October?

A) 14°C B) 21°C C) 23°C D) 32°C

Mean, Median, Mode, & Range Challenge

5. What is the average test score for the class if 6 students received scores of: 78, 82, 97, 66, 73 and 84?

A) 72 B) 78 C) 80 D) 82

6. What is the range of the following numbers?

18, 20, 4, 60, 90, 120, 12

A) 100 B) 116 C) 118 D) 120

7. What is the mode of the following numbers?

11, 11, 13, 15, 15, 20, 25, 32, 40, 32, 60, 70, 32, 70

A) 11 B) 15 C) 32 D) 70

8. If the average of $k + 2$ and $3k + 6$ is m and if the average of $5k$ and $7k - 12$ is n, what is the average of m and n?

A) $2k + 1$ B) $4k - 1$ C) $3k - 1$ D) $5k - 1$

PROBABILITY

Probability: Probability is a fraction or decimal comparing the number of favorable (desired) outcomes to the total number of possible outcomes. It is a number between and including the numbers 0 and 1.

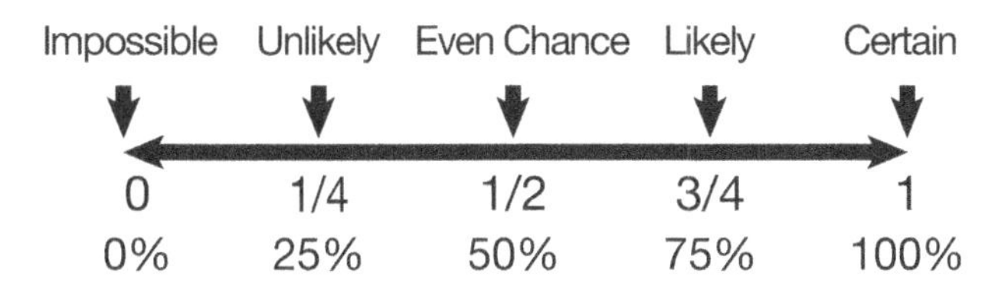

$$\textbf{Probability} = \frac{\text{Number of Desired Outcomes}}{\text{Total Number Of Possible Outcomes}}$$

Example: A letter is chosen at random from the word Mathematics. What is the probability of choosing a?

Solution:

$$\textbf{Probability} = \frac{\text{Number of Desired Outcomes}}{\text{Total Number Of Possible Outcomes}}$$

$$P = \frac{2}{11}$$

Probability Worksheet

The spinner shown is spun once. Find the probability of each event. Write each answer as a fraction, decimal, and percent.

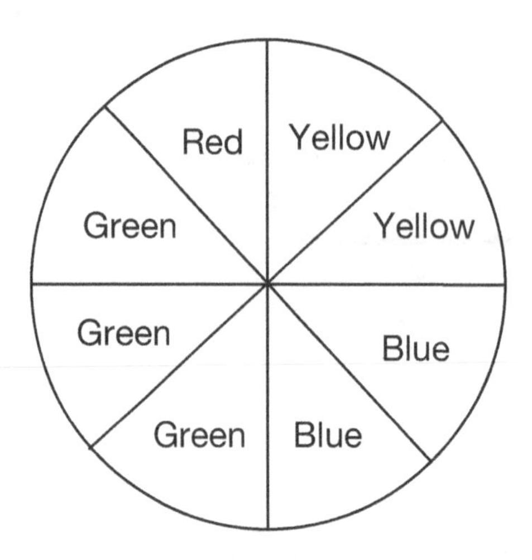

	Fraction	Decimal	Percent
P (blue)			
P (green)			
P (red)			
P (yellow)			

Probability Challenge

1. A bag contains 6 green balls and 4 yellow balls. What is the probability that two balls picked randomly are both of the same color?

A) $\frac{1}{15}$　　B) $\frac{7}{15}$　　C) $\frac{7}{45}$　　D) $\frac{7}{25}$

2. A number from 1 to 12 is chosen at random.
What is the probability of choosing an even number?

A) $\frac{1}{3}$　　B) $\frac{1}{2}$　　C) $\frac{1}{4}$　　D) $\frac{1}{5}$

Use the spinner to answer the question 3 and 4

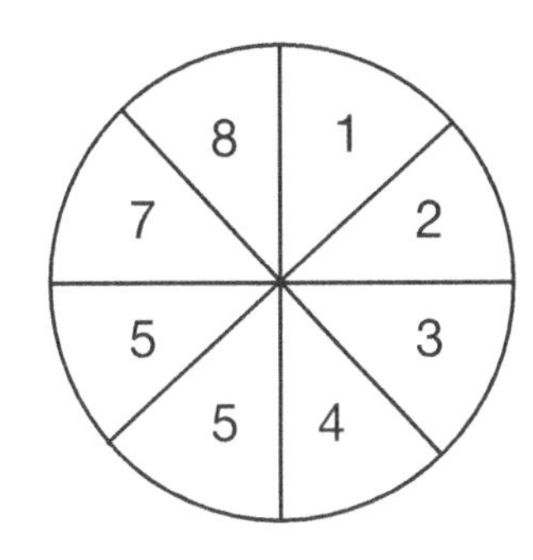

3. From the above spinner, what is the probability of choosing an odd number?

A) $\frac{5}{6}$　　B) $\frac{5}{7}$　　C) $\frac{5}{8}$　　D) $\frac{3}{8}$

4. From the above spinner, what is the probability of choosing an even number?

A) $\frac{4}{13}$　　B) $\frac{5}{13}$　　C) $\frac{5}{8}$　　D) $\frac{3}{8}$

Probability Challenge

5. What measure of central tendency is calculated by the difference between the lowest and highest number of values?

A) Median B) Mean C) Mode D) Range

6. Melisa has 12 green and 7 blue marbles in a bag.
What is the probability of choosing a blue marble from the bag?

A) 7 B) 19 C) $\frac{7}{19}$ D) $\frac{12}{19}$

7. A letter is chosen at random from the word "scientist". What is the probability of choosing either an e or i?

A) $\frac{1}{2}$ B) $\frac{1}{3}$ C) $\frac{1}{4}$ D) $\frac{1}{5}$

8. In a school, the probability that students will choose the math club is $\frac{1}{5}$ and the probability that they will choose the robotics club is $\frac{3}{5}$. If the student can choose only one club, what is the probability of choosing the robotics club or the math club?

A) $\frac{1}{5}$ B) $\frac{3}{5}$ C) $\frac{3}{8}$ D) $\frac{4}{5}$

Number System Worksheet Answer Key

1.

Real Numbers	
Rational Numbers	İrrational Numbers
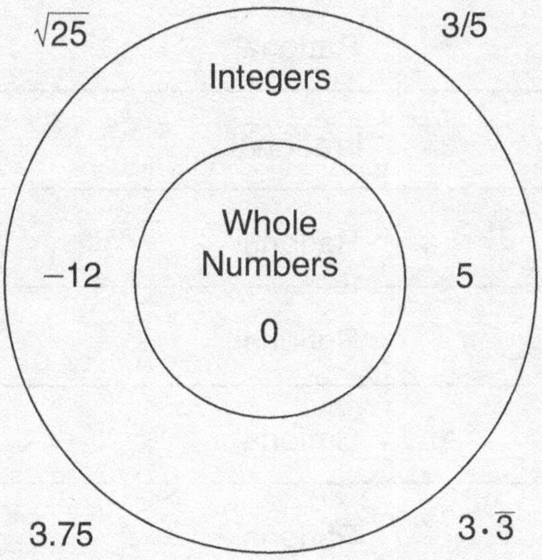 $\sqrt{25}$, 3/5, Integers: −12, 5; Whole Numbers: 0; 3.75, $3.\overline{3}$	π, $\sqrt{10}$, 1.2345 ...

2. A) True

B) True

C) False because an integer cannot be an irrational number.

Number System Worksheet Answer Key

3. Classify each number below as either rational or irrational.

	Rational or Irrational?
0	Rational
$-\frac{1}{2}$	Rational
$\sqrt{20}$	Irrational
$\sqrt{144}$	Rational
$4\frac{1}{3}$	Rational
0.35	Rational
$\overline{9}$	Rational
–9	Rational
4.5	Rational
$\sqrt{7}$	Irrational

4. A) The reciprocal of $\frac{3}{5}$ is $\frac{5}{3}$

B) The reciprocal of -5 is $\frac{-1}{5}$

C) The reciprocal of 10 is $\frac{1}{10}$

D) The reciprocal of 1 is 1

Order of Operations Worksheet Answer Key

1. 22

2. 379

3. 4

4. −49

5. 0

6. 73

7. 33

8. 105

9. 103

10. 16

Prime & Composite Numbers Worksheet Answer Key

1.

1	②	③	4	⑤
6	⑦	8	9	10
⑪	12	⑬	14	15
16	⑰	18	⑲	20
21	22	㉓	24	25

2.

1	2	3	④	5
⑥	7	⑧	⑨	⑩
11	⑫	13	⑭	⑮
⑯	17	⑱	19	⑳
㉑	㉒	23	㉔	㉕

Divisibility Rule Worksheet Answer Key

1. NO, NO, YES

2. YES, YES, NO

3. YES, NO, YES

4. YES, YES, NO

5. YES, NO, NO

6. YES, NO, YES

LCM & GCF Worksheet Answer Key

	GCF	LCM
15, 45	15	45
16, 24	8	48
18, 48	6	144
5, 45	5	45
11, 88	11	88
5, 10, 15	5	30
16, 24, 36	4	144

Absolute Value Worksheet Answer Key

1. 5	**6.** 17	**11.** >	**16.** >
2. 12	**7.** 2	**12.** >	**17.** =
3. 24	**8.** 77	**13.** >	**18.** >
4. 97	**9.** 8	**14.** >	**19.** <
5. 106	**10.** 25	**15.** =	**20.** =

Fractions Worksheet Answer Key

1. $\frac{1}{4}$

2. $\frac{47}{28}$

3. $\frac{25}{33}$

4. $\frac{17}{72}$

5. 1

6. $\frac{1}{18}$

7. $\frac{1}{24}$

8. $\frac{9}{16}$

Decimal Numbers Answer Key

1.	**2.**
$3.5 + 4.4 = 7.9$	$8.9 - 4.8 = 4.1$
$13.4 + 5.7 = 19.1$	$18.90 - 3.9 = 15$
$6.85 + 6.78 = 13.63$	$72.8 - 6.86 = 65.94$
$345.56 + 56.8 = 402.36$	$47.45 - 22.65 = 24.8$
$43.70 + 121.567 = 165.267$	$101.567 - 89.568 = 11.999$
3.	**4.**
$3.8 \times 7 = 26.6$	$4.8 \div 4 = 1.2$
$13.8 \times 5 = 69$	$72.6 \div 3 = 24.2$
$3.8 \times 20 = 76$	$124.8 \div 4 = 31.2$
$4.8 \times 7.2 = 34.56$	$36.8 \div 0.8 = 46$
$12.7 \times 6.74 = 85.598$	$24.8 \div 0.04 = 620$

Rounding Numbers Worksheet Answer Key

1. 7,000,000

2. 35,000,000

3. 503,100

4. 7000

5. 2,000,000

6. 380

7. 456,300

8. 1,230

9. 3,548,000

10. 12,000,000

Laws of Exponents Worksheet Answer Key

1. Base: 2

Exponent: 4

2. Base: a

Exponent: 3b

3. Base: $\frac{1}{2}$

Exponent: 3

4. Exponential form: 3^5

5. Exponential form: 6^{-2}

6. Exponential form: a^{4b}

7. 81

8. 216

9. $\frac{16}{81}$

Laws of Radicals Worksheet Answer Key

1. $\frac{3}{4}$

2. $\sqrt{2}$

3. $\frac{\sqrt{15}}{5}$

4. $\frac{2\sqrt{5}}{5}$

5. $\frac{\sqrt{3}}{2}$

6. $2\sqrt{2}$

7. $5\sqrt{5}$

8. $6\sqrt{2}-2\sqrt{3}$

9. 36

10. 0

11. 46

12. $\sqrt{3}$

13. $5\sqrt{3}$

14. $2\sqrt{3}$

Scientific Notation Worksheet Answer Key

1. 1.25×10^{7}

2. 1.23×10^{3}

3. 9.8×10^{-5}

4. 1.45×10^{8}

5. 4.57×10^{-11}

6. 8.69×10^{-11}

7. 9.67×10^{6}

8. 4.57×10^{-3}

9. 2.19×10^{4}

10. 1.5×10^{0}

11. 4×10^{4}

12. 3.45×10^{7}

13. 1.8×10^{0}

14. 4.44×10^{-2}

15. 3.784×10^{0}

16. 8.9×10^{7}

17. 3.67×10^{0}

18. 2.34×10^{5}

Algebraic Expressions Worksheet Answer Key

1. x

2. $6x$

3. $-5x$

4. $1 - n$

5. $17 - 3n$

6. $-9n$

7. $6 - 8v$

8. $6v - 12$

9. $-4v + 25$

10. $-d - 12$

11. $24d + 36$

12. $\frac{1}{2}d - 6$

13. $-\frac{3}{2}x + 12$

14. $3x - 5$

15. $\frac{x^2}{2} - x$

16. $4x + 3$

Equations with Two Variables Worksheet Answer Key

1. $y = 11$

2. $y = 22$

3. $x = 7$

4. $x = 5$

5. $y = 26$

6. $x = -2$

7. $x = \frac{1}{18}$

8. $y = 6$

9. $x = -27$

10. $x = -9$

11. $y = 16$

12. $x = 13$

Solving Equations & Inequalities Worksheet Answer Key

1. $x = 3.75$

2. $x = -2$

3. $x = 18$

4. $x = 6$

5. $x = -4$

6. $x = -11$

7. $x = 4$

8. $x = 3$

9. $x = -1$

10. $x < 12$

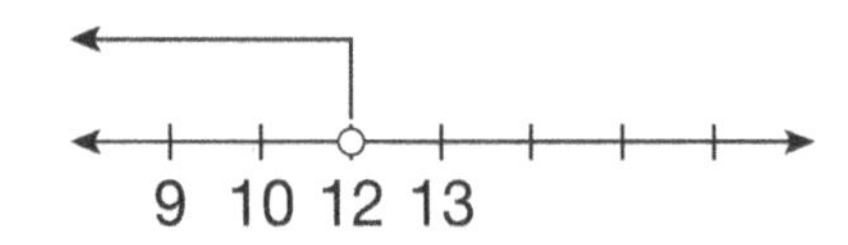

11. $x < -3$

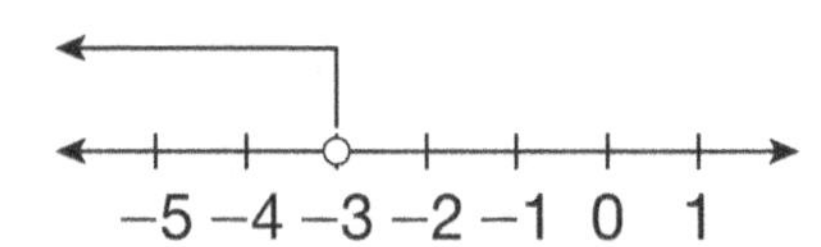

12. $x < -2$

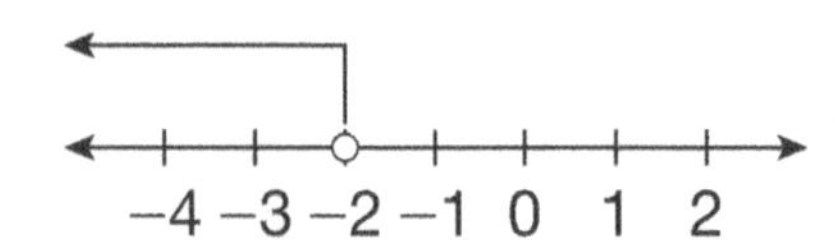

Ratios, Proportional Relations & Variations Worksheet Answer Key

1. Equivalent

2. Equivalent

3. Equivalent

4. Direct

5. Inverse

6. Direct

7. $x = 10$

8. $x = 4.5$

9. $x = 9$

10. $y = 17$

11. $y = 10$

12. $y = 2$

13. 40 boys

Functions Worksheet Answer Key

1. Function: Yes	Domain: (0, 1, 5, 9)	Range: (2, 7, 13, 13)
2. Function: No	Domain: (3, 7, 7, 9)	Range: (4, 5, 8, 19)
3. Function: Yes	Domain: (3, 5, 6, 11)	Range: (2, 4, 7, 8)
4. Function: No	Domain: (a, b, c, a)	Range: (13, 14, 15, 16)

Linear Equations & Slope Worksheet Answer Key

1. $m = \frac{5}{4}$

2. $m = \frac{7}{4}$

3. $m = 2$

4. $m = 3$ and y–intercept $= -24$

5. $m = \frac{1}{3}$ and y–intercept $= -17$

6. $m = \frac{3}{4}$ and y–intercept $= -\frac{1}{4}$

7. $y = 3x - 19$

8. $y = 6x + 9$

9. $m = 3$, y–intercept $= 3$ equation is: $y = 3x + 3$

Unit Rate & Percentages Worksheet Answer Key

1. Unit Rate: 30 miles/day

2. Unit Rate: 4 $/hour

3. Unit Rate: 5 pages/book

4. 40

5. 18

6. 20

7. 9

Angles Worksheet Answer Key

1. $x = 19^{\circ}$

2. $x = 17^{\circ}$

3. $x = 15^{\circ}$

4. $x = 24^{\circ}$

5. $x = 120^{\circ}$

6. $x = 48^{\circ}$

Distance & Midpoint Worksheet Answer Key

1. $d = 5$

2. $d = \sqrt{137}$

3. $d = \sqrt{145}$

4. $d = \frac{1}{6}$

5. $d = 6$

6. $d = \frac{3}{4}$

7. $M = (1,7)$

8. $M = (-5, -2)$

9. $M = (3, 3)$

10. $M = \left(\frac{1}{3}, 9\right)$

11. $M = (0, 4)$

12. $M = (0, 0)$

Triangles and Types of Triangles Worksheet Answer Key

1. Obtuse and scalene triangle

2. Acute and equilateral triangle

3. Acute and isosceles triangle

4. $x = 50°$

5. $y = 35°$

6. $y = 51°$

7. $x = 2$

8. $x = 4\sqrt{3}$

9. $x = 8$

Similarity Theorem Worksheet Answer Key

1. $x = 4$

2. $x = 8$

3. $x = 6$

4. $x = 24$

5. $x = 32$

6. $x = 7.5$

Pythagorean Theorem Worksheet Answer Key

1. The gas station to city hall: $2\sqrt{17}$ units

2. The art museum to school: 8 units

3. The city hall to stadium: $\sqrt{58}$ units

4. The school to stadium: $3\sqrt{2}$ units

5. $x = 10$

6. $x = 15$

7. $x = 4\sqrt{3}$ cm

Coordinate Plane Worksheet Answer Key

Coordinates	Point	Quadrant
(4, 8)	A	I
(−6, 6)	B	II
(3, 3)	C	I
(4, 0)	D	No quadrant
(3, −6)	E	IV
(0, −8)	F	No quadrant
(−8, −5)	G	III

Area & Perimeter Worksheet Answer Key

1. $x = 4cm$

2. $x = 5cm$

3. $x=5cm$

4. $A = 32cm^2$

5. $A = 24cm^2$

6. $A = 22cm^2$

Circles Worksheet Answer Key

1. Diameter: 10cm

   Circumference: 10πcm

   Area: $25\pi cm^2$

2. Diameter: 12ft

   Circumference: 12πft

   Area: $36\pi ft^2$

3. Diameter: 20cm

   Circumference: 20πcm

   Area: $100\pi cm^2$

4. Radius: 6cm

   Circumference: 12πcm

   Area: $36\pi cm^2$

5. Radius: 10ft

   Circumference: 20πft

   Area: $100\pi ft^2$

6. Radius: 15cm

   Circumference: 30πcm

   Area: $225\pi cm^2$

Volume Worksheet Answer Key

1. $V = 343ft^3$

2. $V = 216ft^3$

3. $V = 64ft^3$

4. $V = 102in^3$

5. $V = 210cm^3$

6. $V = 480cm^3$

Mean, Median, Mode, & Range Worksheet Answer Key

1. Mean: 6

   Median: 6

   Mode: 6

   Range: 8

2. Mean: 5

   Median: 6

   Mode: 7

   Range: 6

3. Mean: 0.32

   Median: $\frac{7}{12}$

   Mode: No mode

   Range: $\frac{3}{10}$

4. $y = 14$

5. $z = 28$

6. $k = 48.5$

7. 8

8. Range

Probability Worksheet Answer Key

	Fraction	Decimal	Percent
P (blue)	$\frac{1}{4}$	0.25	25%
P (green)	$\frac{3}{8}$	0.375	37.5%
P (red)	$\frac{1}{8}$	0.125	12.5%
P (yellow)	$\frac{1}{4}$	0.25	25%

CHALLENGE TESTS ANSWER KEYS

Number System Answer Key

1)	2)	3)	4)	5)	6)	7)	8)	9)
C	D	D	B	B	C	B	B	C

Order of Operations Answer Key

1)	2)	3)	4)	5)	6)	7)	8)	9)	10)
D	C	A	D	B	C	D	B	B	C

Prime & Composite Numbers Answer Key

1)	2)	3)	4)	5)	6)	7)	8)	9)	10)
D	D	B	B	C	B	A	D	A	B

Divisibility Rules Answer Key

1)	2)	3)	4)	5)	6)	7)	8)	9)	10)
B	B	D	D	B	C	D	A	YES	D

Least Common Multiple & Greatest Common Factor Answer Key

1)	2)	3)	4)	5)	6)	7)	8)
C	D	B	A	D	A	C	A

Absolute Value Answer Key

1)	2)	3)	4)	5)	6)	7)	8)
D	B	D	D	B	C	B	D

Fractions & Operations With Fractions Answer Key

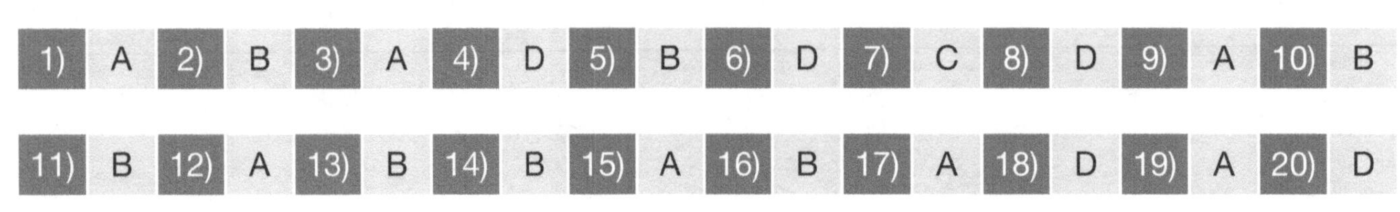

1)	2)	3)	4)	5)	6)	7)	8)	9)	10)
A	B	A	D	B	D	C	D	A	B

11)	12)	13)	14)	15)	16)	17)	18)	19)	20)
B	A	B	B	A	B	A	D	A	D

CHALLENGE TESTS ANSWER KEYS

Decimal Numbers Answer Key

1)	26.4	2)	59.27	3)	97.018	4)	4.86	5)	8.88	6)	28.8611
7)	15.394	8)	16	9)	74.555	10)	5.05	11)	$4.97	12)	20

Rounding Numbers Answer Key

1)	D	2)	B	3)	B	4)	College A and College C	5)	26,000
6)	26,700	7)	26,050	8)	College B	9)	C	10)	D

Laws of Exponents Answer Key

1)	D	2)	A	3)	A	4)	B	5)	B	6)	D	7)	D	8)	D	9)	B	10)	A

Laws of Radicals Answer Key

1)	B	2)	A	3)	B	4)	D	5)	B	6)	D	7)	A	8)	D	9)	D	10)	C

Scientific Notation Answer Key

1)	D	2)	D	3)	D	4)	B	5)	B	6)	C	7)	B	8)	A	9)	D

Algebraic Expressions Answer Key

1)	A	2)	A	3)	B	4)	B	5)	D	6)	B	7)	A	8)	B	9)	C	10)	D

Equations with Two Variables Answer Key

1)	B	2)	C	3)	C	4)	D	5)	D	6)	A	7)	A	8)	B	9)	C

CHALLENGE TESTS ANSWER KEYS

Solving Equations & Inequalities Answer Key

1)	C	2)	D	3)	B	4)	A	5)	B	6)	D	7)	A	8)	A	9)	A	10)	A
11)	A	12)	A																

Ratios, Proportional Relations & Variations Answer Key

1)	D	2)	B	3)	A	4)	C	5)	C	6)	C	7)	A	8)	A	9)	D	10)	B
11)	A	12)	A																

Functions Answer Key

1)	B	2)	D	3)	A	4)	B	5)	C	6)	D	7)	C	8)	D

Linear Equations & Slope Answer Key

1)	C	2)	D	3)	A	4)	C	5)	A	6)	B	7)	D	8)	B	9)	B	10)	A

Unit Rate & Percentages Answer Key

1)	B	2)	D	3)	A	4)	C	5)	B	6)	D	7)	C	8)	D	9)	D	10)	C
11)	C	12)	B																

Angles Answer Key

1)	A	2)	D	3)	B	4)	D	5)	B	6)	C	7)	B	8)	B	9)	D	10)	A

CHALLENGE TESTS ANSWER KEYS

Distance & Midpoint Answer Key

1) C 2) B 3) D 4) C 5) A 6) B 7) C 8) A

Triangles & Type of Triangles Answer Key

1) C 2) D 3) C 4) C 5) D 6) C 7) B 8) B 9) B 10) C

11) B

Similarity Theorem Answer Key

1) C 2) A 3) C 4) D 5) D 6) A 7) A 8) C 9) C 10) C

Pythagorean Theorem Answer Key

1) B 2) B 3) C 4) C 5) A 6) D 7) C 8) B 9) B

Coordinate Plane Answer Key

1) B 2) C 3) D 4) D 5) D 6) D 7) B

Area and Perimeter Answer Key

1) C 2) D 3) C 4) C 5) B 6) B 7) B 8) C 9) C 10) D

Circles, Circumference & Area Answer Key

1) B 2) D 3) A 4) B 5) B 6) D 7) B 8) A

CHALLENGE TESTS ANSWER KEYS

Volume Answer Key

1)	B	2)	C	3)	B	4)	D	5)	B	6)	C	7)	B	8)	B

Mean, Median, Mode & Range Answer Key

1)	D	2)	D	3)	D	4)	C	5)	C	6)	B	7)	C	8)	B

Probability Answer Key

1)	B	2)	B	3)	C	4)	D	5)	D	6)	C	7)	B	8)	D

Made in United States
Cleveland, OH
05 June 2025

17521027R00105